W9-CCH-717

SECOND EDITION

Learning SQL

Alan Beaulieu

O'REILLY®

Beijing · Cambridge · Farnham · Köln · Sebastopol · Taipei · Tokyo

Learning SQL, Second Edition
by Alan Beaulieu

Copyright © 2009 O'Reilly Media, Inc. All rights reserved.
Printed in the United States of America.

Published by O'Reilly Media, Inc., 1005 Gravenstein Highway North, Sebastopol, CA 95472.

O'Reilly books may be purchased for educational, business, or sales promotional use. Online editions are also available for most titles (*http://safari.oreilly.com*). For more information, contact our corporate/institutional sales department: (800) 998-9938 or *corporate@oreilly.com*.

Editor: Mary E. Treseler	**Indexer:** Ellen Troutman Zaig
Production Editor: Loranah Dimant	**Cover Designer:** Karen Montgomery
Copyeditor: Audrey Doyle	**Interior Designer:** David Futato
Proofreader: Nancy Reinhardt	**Illustrator:** Robert Romano

Printing History:

August 2005:	First Edition.
April 2009:	Second Edition.

Nutshell Handbook, the Nutshell Handbook logo, and the O'Reilly logo are registered trademarks of O'Reilly Media, Inc. *Learning SQL*, the image of an Andean marsupial tree frog, and related trade dress are trademarks of O'Reilly Media, Inc.

Many of the designations used by manufacturers and sellers to distinguish their products are claimed as trademarks. Where those designations appear in this book, and O'Reilly Media, Inc. was aware of a trademark claim, the designations have been printed in caps or initial caps.

While every precaution has been taken in the preparation of this book, the publisher and author assume no responsibility for errors or omissions, or for damages resulting from the use of the information contained herein.

RepKover™

This book uses RepKover™, a durable and flexible lay-flat binding.

ISBN: 978-0-596-52083-0

[M]

1239115419

Table of Contents

Preface

Programming languages come and go constantly, and very few languages in use today have roots going back more than a decade or so. Some examples are Cobol, which is still used quite heavily in mainframe environments, and C, which is still quite popular for operating system and server development and for embedded systems. In the database arena, we have SQL, whose roots go all the way back to the 1970s.

SQL is the language for generating, manipulating, and retrieving data from a relational database. One of the reasons for the popularity of relational databases is that properly designed relational databases can handle huge amounts of data. When working with large data sets, SQL is akin to one of those snazzy digital cameras with the high-power zoom lens in that you can use SQL to look at large sets of data, or you can zoom in on individual rows (or anywhere in between). Other database management systems tend to break down under heavy loads because their focus is too narrow (the zoom lens is stuck on maximum), which is why attempts to dethrone relational databases and SQL have largely failed. Therefore, even though SQL is an old language, it is going to be around for a lot longer and has a bright future in store.

Why Learn SQL?

If you are going to work with a relational database, whether you are writing applications, performing administrative tasks, or generating reports, you will need to know how to interact with the data in your database. Even if you are using a tool that generates SQL for you, such as a reporting tool, there may be times when you need to bypass the automatic generation feature and write your own SQL statements.

Learning SQL has the added benefit of forcing you to confront and understand the data structures used to store information about your organization. As you become comfortable with the tables in your database, you may find yourself proposing modifications or additions to your database schema.

Why Use This Book to Do It?

The SQL language is broken into several categories. Statements used to create database objects (tables, indexes, constraints, etc.) are collectively known as SQL *schema statements*. The statements used to create, manipulate, and retrieve the data stored in a database are known as the SQL *data statements*. If you are an administrator, you will be using both SQL schema and SQL data statements. If you are a programmer or report writer, you may only need to use (or be *allowed* to use) SQL data statements. While this book demonstrates many of the SQL schema statements, the main focus of this book is on programming features.

With only a handful of commands, the SQL data statements look deceptively simple. In my opinion, many of the available SQL books help to foster this notion by only skimming the surface of what is possible with the language. However, if you are going to work with SQL, it behooves you to understand fully the capabilities of the language and how different features can be combined to produce powerful results. I feel that this is the only book that provides detailed coverage of the SQL language without the added benefit of doubling as a "door stop" (you know, those 1,250-page "complete references" that tend to gather dust on people's cubicle shelves).

While the examples in this book run on MySQL, Oracle Database, and SQL Server, I had to pick one of those products to host my sample database and to format the result sets returned by the example queries. Of the three, I chose MySQL because it is freely obtainable, easy to install, and simple to administer. For those readers using a different server, I ask that you download and install MySQL and load the sample database so that you can run the examples and experiment with the data.

Structure of This Book

This book is divided into 15 chapters and 3 appendixes:

Chapter 1, *A Little Background*, explores the history of computerized databases, including the rise of the relational model and the SQL language.

Chapter 2, *Creating and Populating a Database*, demonstrates how to create a MySQL database, create the tables used for the examples in this book, and populate the tables with data.

Chapter 3, *Query Primer*, introduces the `select` statement and further demonstrates the most common clauses (`select`, `from`, `where`).

Chapter 4, *Filtering*, demonstrates the different types of conditions that can be used in the `where` clause of a `select`, `update`, or `delete` statement.

Chapter 5, *Querying Multiple Tables*, shows how queries can utilize multiple tables via table joins.

Chapter 6, *Working with Sets*, is all about data sets and how they can interact within queries.

Chapter 7, *Data Generation, Conversion, and Manipulation*, demonstrates several built-in functions used for manipulating or converting data.

Chapter 8, *Grouping and Aggregates*, shows how data can be aggregated.

Chapter 9, *Subqueries*, introduces the subquery (a personal favorite) and shows how and where they can be utilized.

Chapter 10, *Joins Revisited*, further explores the various types of table joins.

Chapter 11, *Conditional Logic*, explores how conditional logic (i.e., if-then-else) can be utilized in `select`, `insert`, `update`, and `delete` statements.

Chapter 12, *Transactions*, introduces transactions and shows how to use them.

Chapter 13, *Indexes and Constraints*, explores indexes and constraints.

Chapter 14, *Views*, shows how to build an interface to shield users from data complexities.

Chapter 15, *Metadata*, demonstrates the utility of the data dictionary.

Appendix A, *ER Diagram for Example Database*, shows the database schema used for all examples in the book.

Appendix B, *MySQL Extensions to the SQL Language*, demonstrates some of the interesting non-ANSI features of MySQL's SQL implementation.

Appendix C, *Solutions to Exercises*, shows solutions to the chapter exercises.

Conventions Used in This Book

The following typographical conventions are used in this book:

Italic
> Used for filenames, directory names, and URLs. Also used for emphasis and to indicate the first use of a technical term.

`Constant width`
> Used for code examples and to indicate SQL keywords within text.

`Constant width italic`
> Used to indicate user-defined terms.

UPPERCASE
> Used to indicate SQL keywords within example code.

`Constant width bold`
> Indicates user input in examples showing an interaction. Also indicates emphasized code elements to which you should pay particular attention.

Indicates a tip, suggestion, or general note. For example, I use notes to point you to useful new features in Oracle9*i*.

 Indicates a warning or caution. For example, I'll tell you if a certain SQL clause might have unintended consequences if not used carefully.

How to Contact Us

Please address comments and questions concerning this book to the publisher:

> O'Reilly Media, Inc.
> 1005 Gravenstein Highway North
> Sebastopol, CA 95472
> 800-998-9938 (in the United States or Canada)
> 707-829-0515 (international or local)
> 707-829-0104 (fax)

O'Reilly maintains a web page for this book, which lists errata, examples, and any additional information. You can access this page at:

> *http://www.oreilly.com/catalog/9780596520830*

To comment or ask technical questions about this book, send email to:

> *bookquestions@oreilly.com*

For more information about O'Reilly books, conferences, Resource Centers, and the O'Reilly Network, see the website at:

> *http://www.oreilly.com*

Using Code Examples

This book is here to help you get your job done. In general, you may use the code in this book in your programs and documentation. You do not need to contact us for permission unless you're reproducing a significant portion of the code. For example, writing a program that uses several chunks of code from this book does not require permission. Selling or distributing a CD-ROM of examples from O'Reilly books does require permission. Answering a question by citing this book and quoting example code does not require permission. Incorporating a significant amount of example code from this book into your product's documentation does require permission.

We appreciate, but do not require, attribution. An attribution usually includes the title, author, publisher, and ISBN. For example, "*Learning SQL*, Second Edition, by Alan Beaulieu. Copyright 2009 O'Reilly Media, Inc., 978-0-596-52083-0."

If you feel your use of code examples falls outside fair use or the permission given above, feel free to contact us at *permissions@oreilly.com*.

Safari® Books Online

When you see a Safari® Books Online icon on the cover of your favorite technology book, that means the book is available online through the O'Reilly Network Safari Bookshelf.

Safari offers a solution that's better than e-books. It's a virtual library that lets you easily search thousands of top tech books, cut and paste code samples, download chapters, and find quick answers when you need the most accurate, current information. Try it for free at *http://my.safaribooksonline.com*.

Acknowledgments

I would like to thank my editor, Mary Treseler, for helping to make this second edition a reality, and many thanks to Kevin Kline, Roy Owens, Richard Sonen, and Matthew Russell, who were kind enough to review the book for me over the Christmas/New Year holidays. I would also like to thank the many readers of my first edition who were kind enough to send questions, comments, and corrections. Lastly, I thank my wife, Nancy, and my daughters, Michelle and Nicole, for their encouragement and inspiration.

CHAPTER 1

A Little Background

Before we roll up our sleeves and get to work, it might be beneficial to introduce some basic database concepts and look at the history of computerized data storage and retrieval.

Introduction to Databases

A *database* is nothing more than a set of related information. A telephone book, for example, is a database of the names, phone numbers, and addresses of all people living in a particular region. While a telephone book is certainly a ubiquitous and frequently used database, it suffers from the following:

- Finding a person's telephone number can be time-consuming, especially if the telephone book contains a large number of entries.
- A telephone book is indexed only by last/first names, so finding the names of the people living at a particular address, while possible in theory, is not a practical use for this database.
- From the moment the telephone book is printed, the information becomes less and less accurate as people move into or out of a region, change their telephone numbers, or move to another location within the same region.

The same drawbacks attributed to telephone books can also apply to any manual data storage system, such as patient records stored in a filing cabinet. Because of the cumbersome nature of paper databases, some of the first computer applications developed were *database systems*, which are computerized data storage and retrieval mechanisms. Because a database system stores data electronically rather than on paper, a database system is able to retrieve data more quickly, index data in multiple ways, and deliver up-to-the-minute information to its user community.

Early database systems managed data stored on magnetic tapes. Because there were generally far more tapes than tape readers, technicians were tasked with loading and unloading tapes as specific data was requested. Because the computers of that era had very little memory, multiple requests for the same data generally required the data to be read from the tape multiple times. While these database systems were a significant improvement over paper databases, they are a far cry from what is possible with today's technology. (Modern database systems can manage terabytes of data spread across many fast-access disk drives, holding tens of gigabytes of that data in high-speed memory, but I'm getting a bit ahead of myself.)

Nonrelational Database Systems

 This section contains some background information about prerelational database systems. For those readers eager to dive into SQL, feel free to skip ahead a couple of pages to the next section.

Over the first several decades of computerized database systems, data was stored and represented to users in various ways. In a *hierarchical database system*, for example, data is represented as one or more tree structures. Figure 1-1 shows how data relating to George Blake's and Sue Smith's bank accounts might be represented via tree structures.

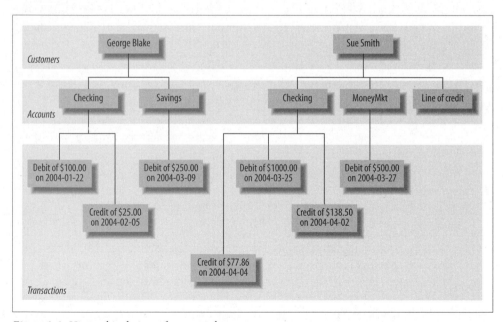

Figure 1-1. Hierarchical view of account data

George and Sue each have their own tree containing their accounts and the transactions on those accounts. The hierarchical database system provides tools for locating a particular customer's tree and then traversing the tree to find the desired accounts and/or transactions. Each node in the tree may have either zero or one parent and zero, one, or many children. This configuration is known as a *single-parent hierarchy*.

Another common approach, called the *network database system*, exposes sets of records and sets of links that define relationships between different records. Figure 1-2 shows how George's and Sue's same accounts might look in such a system.

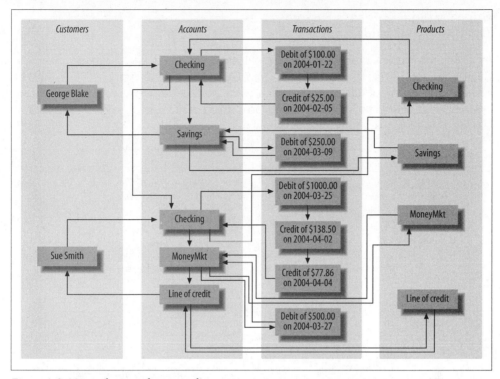

Figure 1-2. Network view of account data

In order to find the transactions posted to Sue's money market account, you would need to perform the following steps:

1. Find the customer record for Sue Smith.
2. Follow the link from Sue Smith's customer record to her list of accounts.
3. Traverse the chain of accounts until you find the money market account.
4. Follow the link from the money market record to its list of transactions.

One interesting feature of network database systems is demonstrated by the set of `product` records on the far right of Figure 1-2. Notice that each `product` record (Checking, Savings, etc.) points to a list of `account` records that are of that product type. `Account` records, therefore, can be accessed from multiple places (both `customer` records and `product` records), allowing a network database to act as a *multiparent hierarchy*.

Both hierarchical and network database systems are alive and well today, although generally in the mainframe world. Additionally, hierarchical database systems have enjoyed a rebirth in the directory services realm, such as Microsoft's Active Directory and the Red Hat Directory Server, as well as with Extensible Markup Language (XML). Beginning in the 1970s, however, a new way of representing data began to take root, one that was more rigorous yet easy to understand and implement.

The Relational Model

In 1970, Dr. E. F. Codd of IBM's research laboratory published a paper titled "A Relational Model of Data for Large Shared Data Banks" that proposed that data be represented as sets of *tables*. Rather than using pointers to navigate between related entities, redundant data is used to link records in different tables. Figure 1-3 shows how George's and Sue's account information would appear in this context.

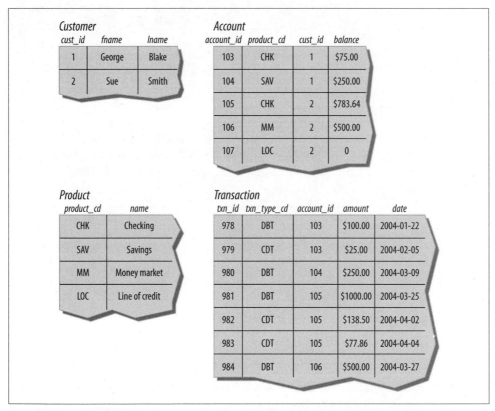

Figure 1-3. Relational view of account data

There are four tables in Figure 1-3 representing the four entities discussed so far: customer, product, account, and transaction. Looking across the top of the customer table in Figure 1-3, you can see three *columns*: `cust_id` (which contains the customer's ID number), `fname` (which contains the customer's first name), and `lname` (which contains the customer's last name). Looking down the side of the `customer` table, you can see two *rows*, one containing George Blake's data and the other containing Sue Smith's data. The number of columns that a table may contain differs from server to server, but it is generally large enough not to be an issue (Microsoft SQL Server, for example, allows up to 1,024 columns per table). The number of rows that a table may contain is more a matter of physical limits (i.e., how much disk drive space is available) and maintainability (i.e., how large a table can get before it becomes difficult to work with) than of database server limitations.

Each table in a relational database includes information that uniquely identifies a row in that table (known as the *primary key*), along with additional information needed to describe the entity completely. Looking again at the `customer` table, the `cust_id` column holds a different number for each customer; George Blake, for example, can be uniquely identified by customer ID #1. No other customer will ever be assigned that identifier, and no other information is needed to locate George Blake's data in the `customer` table.

Every database server provides a mechanism for generating unique sets of numbers to use as primary key values, so you won't need to worry about keeping track of what numbers have been assigned.

While I might have chosen to use the combination of the `fname` and `lname` columns as the primary key (a primary key consisting of two or more columns is known as a *compound key*), there could easily be two or more people with the same first and last names that have accounts at the bank. Therefore, I chose to include the `cust_id` column in the `customer` table specifically for use as a primary key column.

In this example, choosing `fname`/`lname` as the primary key would be referred to as a *natural key*, whereas the choice of `cust_id` would be referred to as a *surrogate key*. The decision whether to employ natural or surrogate keys is a topic of widespread debate, but in this particular case the choice is clear, since a person's last name may change (such as when a person adopts a spouse's last name), and primary key columns should never be allowed to change once a value has been assigned.

Some of the tables also include information used to navigate to another table; this is where the "redundant data" mentioned earlier comes in. For example, the `account` table includes a column called `cust_id`, which contains the unique identifier of the customer who opened the account, along with a column called `product_cd`, which contains the unique identifier of the product to which the account will conform. These columns are known as *foreign keys*, and they serve the same purpose as the lines that connect the entities in the hierarchical and network versions of the account information. If you are looking at a particular account record and want to know more information about the customer who opened the account, you would take the value of the `cust_id` column and use it to find the appropriate row in the `customer` table (this process is known, in relational database lingo, as a *join*; joins are introduced in Chapter 3 and probed deeply in Chapters 5 and 10).

It might seem wasteful to store the same data many times, but the relational model is quite clear on what redundant data may be stored. For example, it is proper for the `account` table to include a column for the unique identifier of the customer who opened the account, but it is not proper to include the customer's first and last names in the `account` table as well. If a customer were to change her name, for example, you want to make sure that there is only one place in the database that holds the customer's name; otherwise, the data might be changed in one place but not another, causing the data in the database to be unreliable. The proper place for this data is the `customer` table, and only the `cust_id` values should be included in other tables. It is also not proper for a single column to contain multiple pieces of information, such as a `name` column that contains both a person's first and last names, or an `address` column that contains street, city, state, and zip code information. The process of refining a database design to ensure that each independent piece of information is in only one place (except for foreign keys) is known as *normalization*.

Getting back to the four tables in Figure 1-3, you may wonder how you would use these tables to find George Blake's transactions against his checking account. First, you would find George Blake's unique identifier in the `customer` table. Then, you would find the row in the `account` table whose `cust_id` column contains George's unique identifier and whose `product_cd` column matches the row in the `product` table whose `name` column equals "Checking." Finally, you would locate the rows in the `transaction` table whose `account_id` column matches the unique identifier from the `account` table. This might sound complicated, but you can do it in a single command, using the SQL language, as you will see shortly.

Some Terminology

I introduced some new terminology in the previous sections, so maybe it's time for some formal definitions. Table 1-1 shows the terms we use for the remainder of the book along with their definitions.

Table 1-1. Terms and definitions

Term	Definition
Entity	Something of interest to the database user community. Examples include customers, parts, geographic locations, etc.
Column	An individual piece of data stored in a table.
Row	A set of columns that together completely describe an entity or some action on an entity. Also called a record.
Table	A set of rows, held either in memory (nonpersistent) or on permanent storage (persistent).

Term	Definition
Result set	Another name for a nonpersistent table, generally the result of an SQL query.
Primary key	One or more columns that can be used as a unique identifier for each row in a table.
Foreign key	One or more columns that can be used together to identify a single row in another table.

What Is SQL?

Along with Codd's definition of the relational model, he proposed a language called DSL/Alpha for manipulating the data in relational tables. Shortly after Codd's paper was released, IBM commissioned a group to build a prototype based on Codd's ideas. This group created a simplified version of DSL/Alpha that they called SQUARE. Refinements to SQUARE led to a language called SEQUEL, which was, finally, renamed SQL.

SQL is now entering middle age (as is this author, alas), and it has undergone a great deal of change along the way. In the mid-1980s, the American National Standards Institute (ANSI) began working on the first standard for the SQL language, which was published in 1986. Subsequent refinements led to new releases of the SQL standard in 1989, 1992, 1999, 2003, and 2006. Along with refinements to the core language, new features have been added to the SQL language to incorporate object-oriented functionality, among other things. The latest standard, SQL:2006, focuses on the integration of SQL and XML and defines a language called XQuery which is used to query data in XML documents.

SQL goes hand in hand with the relational model because the result of an SQL query is a table (also called, in this context, a *result set*). Thus, a new permanent table can be created in a relational database simply by storing the result set of a query. Similarly, a query can use both permanent tables and the result sets from other queries as inputs (we explore this in detail in Chapter 9).

One final note: SQL is not an acronym for anything (although many people will insist it stands for "Structured Query Language"). When referring to the language, it is equally acceptable to say the letters individually (i.e., S. Q. L.) or to use the word *sequel*.

SQL Statement Classes

The SQL language is divided into several distinct parts: the parts that we explore in this book include *SQL schema statements*, which are used to define the data structures stored in the database; *SQL data statements*, which are used to manipulate the data structures previously defined using SQL schema statements; and *SQL transaction statements*, which are used to begin, end, and roll back transactions (covered in Chapter 12). For example, to create a new table in your database, you would use the SQL schema statement `create table`, whereas the process of populating your new table with data would require the SQL data statement `insert`.

To give you a taste of what these statements look like, here's an SQL schema statement that creates a table called `corporation`:

```
CREATE TABLE corporation
 (corp_id SMALLINT,
  name VARCHAR(30),
  CONSTRAINT pk_corporation PRIMARY KEY (corp_id)
 );
```

This statement creates a table with two columns, `corp_id` and `name`, with the `corp_id` column identified as the primary key for the table. We probe the finer details of this statement, such as the different data types available with MySQL, in Chapter 2. Next, here's an SQL data statement that inserts a row into the `corporation` table for Acme Paper Corporation:

```
INSERT INTO corporation (corp_id, name)
VALUES (27, 'Acme Paper Corporation');
```

This statement adds a row to the `corporation` table with a value of 27 for the `corp_id` column and a value of `Acme Paper Corporation` for the `name` column.

Finally, here's a simple `select` statement to retrieve the data that was just created:

```
mysql< SELECT name
    -> FROM corporation
    -> WHERE corp_id = 27;
+------------------------+
| name                   |
+------------------------+
| Acme Paper Corporation |
+------------------------+
```

All database elements created via SQL schema statements are stored in a special set of tables called the *data dictionary*. This "data about the database" is known collectively as *metadata* and is explored in Chapter 15. Just like tables that you create yourself, data dictionary tables can be queried via a `select` statement, thereby allowing you to discover the current data structures deployed in the database at runtime. For example, if you are asked to write a report showing the new accounts created last month, you could either hardcode the names of the columns in the `account` table that were known to you when you wrote the report, or query the data dictionary to determine the current set of columns and dynamically generate the report each time it is executed.

Most of this book is concerned with the data portion of the SQL language, which consists of the `select`, `update`, `insert`, and `delete` commands. SQL schema statements is demonstrated in Chapter 2, where the sample database used throughout this book is generated. In general, SQL schema statements do not require much discussion apart from their syntax, whereas SQL data statements, while few in number, offer numerous opportunities for detailed study. Therefore, while I try to introduce you to many of the SQL schema statements, most chapters in this book concentrate on the SQL data statements.

SQL: A Nonprocedural Language

If you have worked with programming languages in the past, you are used to defining variables and data structures, using conditional logic (i.e., if-then-else) and looping constructs (i.e., do while ... end), and breaking your code into small, reusable pieces (i.e., objects, functions, procedures). Your code is handed to a compiler, and the executable that results does exactly (well, not always *exactly*) what you programmed it to do. Whether you work with Java, C#, C, Visual Basic, or some other *procedural* language, you are in complete control of what the program does.

A procedural language defines both the desired results and the mechanism, or process, by which the results are generated. Nonprocedural languages also define the desired results, but the process by which the results are generated is left to an external agent.

With SQL, however, you will need to give up some of the control you are used to, because SQL statements define the necessary inputs and outputs, but the manner in which a statement is executed is left to a component of your database engine known as the *optimizer*. The optimizer's job is to look at your SQL statements and, taking into account how your tables are configured and what indexes are available, decide the most efficient execution path (well, not always the *most* efficient). Most database engines will allow you to influence the optimizer's decisions by specifying *optimizer hints*, such as suggesting that a particular index be used; most SQL users, however, will never get to this level of sophistication and will leave such tweaking to their database administrator or performance expert.

With SQL, therefore, you will not be able to write complete applications. Unless you are writing a simple script to manipulate certain data, you will need to integrate SQL with your favorite programming language. Some database vendors have done this for you, such as Oracle's PL/SQL language, MySQL's stored procedure language, and Microsoft's Transact-SQL language. With these languages, the SQL data statements are part of the language's grammar, allowing you to seamlessly integrate database queries with procedural commands. If you are using a non-database-specific language such as Java, however, you will need to use a toolkit/API to execute SQL statements from your code. Some of these toolkits are provided by your database vendor, whereas others are created by third-party vendors or by open source providers. Table 1-2 shows some of the available options for integrating SQL into a specific language.

Table 1-2. SQL integration toolkits

Language	Toolkit
Java	JDBC (Java Database Connectivity; JavaSoft)
C++	Rogue Wave SourcePro DB (third-party tool to connect to Oracle, SQL Server, MySQL, Informix, DB2, Sybase, and PostgreSQL databases)
C/C++	Pro*C (Oracle), MySQL C API (open source), and DB2 Call Level Interface (IBM)
C#	ADO.NET (Microsoft)
Perl	Perl DBI
Python	Python DB
Visual Basic	ADO.NET (Microsoft)

If you only need to execute SQL commands interactively, every database vendor provides at least a simple command-line tool for submitting SQL commands to the database engine and inspecting the results. Most vendors provide a graphical tool as well that includes one window showing your SQL commands and another window showing the results from your SQL commands. Since the examples in this book are executed against a MySQL database, I use the `mysql` command-line tool that is included as part of the MySQL installation to run the examples and format the results.

SQL Examples

Earlier in this chapter, I promised to show you an SQL statement that would return all the transactions against George Blake's checking account. Without further ado, here it is:

```
SELECT t.txn_id, t.txn_type_cd, t.txn_date, t.amount
FROM individual i
  INNER JOIN account a ON i.cust_id = a.cust_id
  INNER JOIN product p ON p.product_cd = a.product_cd
  INNER JOIN transaction t ON t.account_id = a.account_id
WHERE i.fname = 'George' AND i.lname = 'Blake'
  AND p.name = 'checking account';

+--------+-------------+---------------------+--------+
| txn_id | txn_type_cd | txn_date            | amount |
+--------+-------------+---------------------+--------+
|     11 | DBT         | 2008-01-05 00:00:00 | 100.00 |
+--------+-------------+---------------------+--------+
1 row in set (0.00 sec)
```

Without going into too much detail at this point, this query identifies the row in the `individual` table for George Blake and the row in the `product` table for the "checking" product, finds the row in the `account` table for this individual/product combination, and returns four columns from the `transaction` table for all transactions posted to this account. If you happen to know that George Blake's customer ID is 8 and that checking accounts are designated by the code `'CHK'`, then you can simply find George Blake's checking account in the `account` table based on the customer ID and use the account ID to find the appropriate transactions:

```
SELECT t.txn_id, t.txn_type_cd, t.txn_date, t.amount
FROM account a
  INNER JOIN transaction t ON t.account_id = a.account_id
WHERE a.cust_id = 8 AND a.product_cd = 'CHK';
```

I cover all of the concepts in these queries (plus a lot more) in the following chapters, but I wanted to at least show what they would look like.

The previous queries contain three different *clauses*: `select`, `from`, and `where`. Almost every query that you encounter will include at least these three clauses, although there are several more that can be used for more specialized purposes. The role of each of these three clauses is demonstrated by the following:

```
SELECT /* one or more things */ ...
FROM /* one or more places */ ...
WHERE /* one or more conditions apply */ ...
```

 Most SQL implementations treat any text between the /* and */ tags as comments.

When constructing your query, your first task is generally to determine which table or tables will be needed and then add them to your `from` clause. Next, you will need to add conditions to your `where` clause to filter out the data from these tables that you aren't interested in. Finally, you will decide which columns from the different tables need to be retrieved and add them to your `select` clause. Here's a simple example that shows how you would find all customers with the last name "Smith":

```
SELECT cust_id, fname
FROM individual
WHERE lname = 'Smith';
```

This query searches the `individual` table for all rows whose `lname` column matches the string `'Smith'` and returns the `cust_id` and `fname` columns from those rows.

Along with querying your database, you will most likely be involved with populating and modifying the data in your database. Here's a simple example of how you would insert a new row into the `product` table:

```
INSERT INTO product (product_cd, name)
VALUES ('CD', 'Certificate of Depysit')
```

Whoops, looks like you misspelled "Deposit." No problem. You can clean that up with an `update` statement:

```
UPDATE product
SET name = 'Certificate of Deposit'
WHERE product_cd = 'CD';
```

Notice that the update statement also contains a where clause, just like the select statement. This is because an update statement must identify the rows to be modified; in this case, you are specifying that only those rows whose product_cd column matches the string 'CD' should be modified. Since the product_cd column is the primary key for the product table, you should expect your update statement to modify exactly one row (or zero, if the value doesn't exist in the table). Whenever you execute an SQL data statement, you will receive feedback from the database engine as to how many rows were affected by your statement. If you are using an interactive tool such as the mysql command-line tool mentioned earlier, then you will receive feedback concerning how many rows were either:

- Returned by your select statement
- Created by your insert statement
- Modified by your update statement
- Removed by your delete statement

If you are using a procedural language with one of the toolkits mentioned earlier, the toolkit will include a call to ask for this information after your SQL data statement has executed. In general, it's a good idea to check this info to make sure your statement didn't do something unexpected (like when you forget to put a where clause on your delete statement and delete every row in the table!).

What Is MySQL?

Relational databases have been available commercially for over two decades. Some of the most mature and popular commercial products include:

- Oracle Database from Oracle Corporation
- SQL Server from Microsoft
- DB2 Universal Database from IBM
- Sybase Adaptive Server from Sybase

All these database servers do approximately the same thing, although some are better equipped to run very large or very-high-throughput databases. Others are better at handling objects or very large files or XML documents, and so on. Additionally, all these servers do a pretty good job of complying with the latest ANSI SQL standard. This is a good thing, and I make it a point to show you how to write SQL statements that will run on any of these platforms with little or no modification.

Along with the commercial database servers, there has been quite a bit of activity in the open source community in the past five years with the goal of creating a viable alternative to the commercial database servers. Two of the most commonly used open source database servers are PostgreSQL and MySQL. The MySQL website (*http://www .mysql.com*) currently claims over 10 million installations, its server is available for free, and I have found its server to be extremely simple to download and install. For these reasons, I have decided that all examples for this book be run against a MySQL (version 6.0) database, and that the `mysql` command-line tool be used to format query results. Even if you are already using another server and never plan to use MySQL, I urge you to install the latest MySQL server, load the sample schema and data, and experiment with the data and examples in this book.

However, keep in mind the following caveat:

This is not a book about MySQL's SQL implementation.

Rather, this book is designed to teach you how to craft SQL statements that will run on MySQL with no modifications, and will run on recent releases of Oracle Database, Sybase Adaptive Server, and SQL Server with few or no modifications.

To keep the code in this book as vendor-independent as possible, I will refrain from demonstrating some of the interesting things that the MySQL SQL language implementers have decided to do that can't be done on other database implementations. Instead, Appendix B covers some of these features for readers who are planning to continue using MySQL.

What's in Store

The overall goal of the next four chapters is to introduce the SQL data statements, with a special emphasis on the three main clauses of the `select` statement. Additionally, you will see many examples that use the bank schema (introduced in the next chapter), which will be used for all examples in the book. It is my hope that familiarity with a single database will allow you to get to the crux of an example without your having to stop and examine the tables being used each time. If it becomes a bit tedious working with the same set of tables, feel free to augment the sample database with additional tables, or invent your own database with which to experiment.

After you have a solid grasp on the basics, the remaining chapters will drill deep into additional concepts, most of which are independent of each other. Thus, if you find yourself getting confused, you can always move ahead and come back later to revisit a chapter. When you have finished the book and worked through all of the examples, you will be well on your way to becoming a seasoned SQL practitioner.

For readers interested in learning more about relational databases, the history of computerized database systems, or the SQL language than was covered in this short introduction, here are a few resources worth checking out:

- C.J. Date's *Database in Depth: Relational Theory for Practitioners* (*http://oreilly .com/catalog/9780596100124/*) (O'Reilly)
- C.J. Date's *An Introduction to Database Systems*, Eighth Edition (Addison-Wesley)

- C.J. Date's *The Database Relational Model: A Retrospective Review and Analysis: A Historical Account and Assessment of E. F. Codd's Contribution to the Field of Database Technology* (Addison-Wesley)
- *http://en.wikipedia.org/wiki/Database_management_system*
- *http://www.mcjones.org/System_R/*

CHAPTER 2

Creating and Populating a Database

This chapter provides you with the information you need to create your first database and to create the tables and associated data used for the examples in this book. You will also learn about various data types and see how to create tables using them. Because the examples in this book are executed against a MySQL database, this chapter is somewhat skewed toward MySQL's features and syntax, but most concepts are applicable to any server.

Creating a MySQL Database

If you already have a MySQL database server available for your use, you can skip the installation instructions and start with the instructions in Table 2-1. Keep in mind, however, that this book assumes that you are using MySQL version 6.0 or later, so you may want to consider upgrading your server or installing another server if you are using an earlier release.

The following instructions show you the minimum steps required to install a MySQL 6.0 server on a Windows computer:

1. Go to the download page for the MySQL Database Server at *http://dev.mysql.com/ downloads*. If you are loading version 6.0, the full URL is *http://dev.mysql.com/ downloads/mysql/6.0.html*.

2. Download the Windows Essentials (x86) package, which includes only the commonly used tools.

3. When asked "Do you want to run or save this file?" click Run.

4. The MySQL Server 6.0—Setup Wizard window appears. Click Next.

5. Activate the Typical Install radio button, and click Next.

6. Click Install.

7. A MySQL Enterprise window appears. Click Next twice.

8. When the installation is complete, make sure the box is checked next to "Configure the MySQL Server now," and then click Finish. This launches the Configuration Wizard.

9. When the Configuration Wizard launches, activate the Standard Configuration radio button, and then select both the "Install as Windows Service" and "Include Bin Directory in Windows Path" checkboxes. Click Next.

10. Select the Modify Security Settings checkbox and enter a password for the root user (make sure you write down the password, because you will need it shortly!), and click Next.

11. Click Execute.

At this point, if all went well, the MySQL server is installed and running. If not, I suggest you uninstall the server and read the "Troubleshooting a MySQL Installation Under Windows" guide (which you can find at *http://dev.mysql.com/doc/refman/6.0/en/windows-troubleshooting.html*).

 If you uninstalled an older version of MySQL before loading version 6.0, you may have some further cleanup to do (I had to clean out some old Registry entries) before you can get the Configuration Wizard to run successfully.

Next, you will need to open a Windows command window, launch the mysql tool, and create your database and database user. Table 2-1 describes the necessary steps. In step 5, feel free to choose your own password for the lrngsql user rather than "xyz" (but don't forget to write it down!).

Table 2-1. Creating the sample database

Step	Description	Action
1	Open the Run dialog box from the Start menu	Choose Start and then Run
2	Launch a command window	Type **cmd** and click OK
3	Log in to MySQL as root	mysql -u root -p
4	Create a database for the sample data	create database bank;
5	Create the lrngsql database user with full privileges on the bank database	grant all privileges on bank.* to 'lrngsql'@'localhost' identified by 'xyz';
6	Exit the mysql tool	quit;
7	Log in to MySQL as lrngsql	mysql -u lrngsql -p;
8	Attach to the bank database	use bank;

You now have a MySQL server, a database, and a database user; the only thing left to do is create the database tables and populate them with sample data. To do so, download the script at *http://examples.oreilly.com/learningsql/* and run it from the `mysql` utility. If you saved the file as *c:\temp\LearningSQLExample.sql*, you would need to do the following:

1. If you have logged out of the `mysql` tool, repeat steps 7 and 8 from Table 2-1.

2. Type **`source c:\temp\LearningSQLExample.sql;`** and press Enter.

You should now have a working database populated with all the data needed for the examples in this book.

Using the mysql Command-Line Tool

Whenever you invoke the `mysql` command-line tool, you can specify the username and database to use, as in the following:

```
mysql -u lrngsql -p bank
```

This will save you from having to type `use bank;` every time you start up the tool. You will be asked for your password, and then the `mysql>` prompt will appear, via which you will be able to issue SQL statements and view the results. For example, if you want to know the current date and time, you could issue the following query:

```
mysql> SELECT now();
+---------------------+
| now()               |
+---------------------+
| 2008-02-19 16:48:46 |
+---------------------+
1 row in set (0.01 sec)
```

The `now()` function is a built-in MySQL function that returns the current date and time. As you can see, the `mysql` command-line tool formats the results of your queries within a rectangle bounded by +, -, and | characters. After the results have been exhausted (in this case, there is only a single row of results), the `mysql` command-line tool shows how many rows were returned and how long the SQL statement took to execute.

About Missing from Clauses

With some database servers, you won't be able to issue a query without a `from` clause that names at least one table. Oracle Database is a commonly used server for which this is true. For cases when you only need to call a function, Oracle provides a table called `dual`, which consists of a single column called `dummy` that contains a single row of data. In order to be compatible with Oracle Database, MySQL also provides a `dual` table. The previous query to determine the current date and time could therefore be written as:

```
mysql> SELECT now()
    -> FROM dual;
+---------------------+
| now()               |
+---------------------+
| 2005-05-06 16:48:46 |
+---------------------+
1 row in set (0.01 sec)
```

If you are not using Oracle and have no need to be compatible with Oracle, you can ignore the `dual` table altogether and use just a `select` clause without a `from` clause.

When you are done with the `mysql` command-line tool, simply type **quit;** or **exit;** to return to the Windows command shell.

MySQL Data Types

In general, all the popular database servers have the capacity to store the same types of data, such as strings, dates, and numbers. Where they typically differ is in the specialty data types, such as XML documents or very large text or binary documents. Since this is an introductory book on SQL, and since 98% of the columns you encounter will be simple data types, this book covers only the character, date, and numeric data types.

Character Data

Character data can be stored as either fixed-length or variable-length strings; the difference is that fixed-length strings are right-padded with spaces and always consume the same number of bytes, and variable-length strings are not right-padded with spaces and don't always consume the same number of bytes. When defining a character column, you must specify the maximum size of any string to be stored in the column. For example, if you want to store strings up to 20 characters in length, you could use either of the following definitions:

```
char(20)    /* fixed-length */
varchar(20) /* variable-length */
```

The maximum length for `char` columns is currently 255 bytes, whereas `varchar` columns can be up to 65,535 bytes. If you need to store longer strings (such as emails, XML documents, etc.), then you will want to use one of the text types (`mediumtext` and `longtext`), which I cover later in this section. In general, you should use the `char` type when all strings to be stored in the column are of the same length, such as state abbreviations, and the `varchar` type when strings to be stored in the column are of varying lengths. Both `char` and `varchar` are used in a similar fashion in all the major database servers.

 Oracle Database is an exception when it comes to the use of `varchar`. Oracle users should use the `varchar2` type when defining variable-length character columns.

Character sets

For languages that use the Latin alphabet, such as English, there is a sufficiently small number of characters such that only a single byte is needed to store each character. Other languages, such as Japanese and Korean, contain large numbers of characters, thus requiring multiple bytes of storage for each character. Such character sets are therefore called *multibyte character sets*.

MySQL can store data using various character sets, both single- and multibyte. To view the supported character sets in your server, you can use the `show` command, as in:

```
mysql> SHOW CHARACTER SET;
+----------+-----------------------------+---------------------+--------+
| Charset  | Description                 | Default collation   | Maxlen |
+----------+-----------------------------+---------------------+--------+
| big5     | Big5 Traditional Chinese    | big5_chinese_ci     |      2 |
| dec8     | DEC West European           | dec8_swedish_ci     |      1 |
| cp850    | DOS West European           | cp850_general_ci    |      1 |
| hp8      | HP West European            | hp8_english_ci      |      1 |
| koi8r    | KOI8-R Relcom Russian       | koi8r_general_ci    |      1 |
| latin1   | cp1252 West European        | latin1_swedish_ci   |      1 |
| latin2   | ISO 8859-2 Central European | latin2_general_ci   |      1 |
| swe7     | 7bit Swedish                | swe7_swedish_ci     |      1 |
| ascii    | US ASCII                    | ascii_general_ci    |      1 |
| ujis     | EUC-JP Japanese             | ujis_japanese_ci    |      3 |
| sjis     | Shift-JIS Japanese          | sjis_japanese_ci    |      2 |
| hebrew   | ISO 8859-8 Hebrew           | hebrew_general_ci   |      1 |
| tis620   | TIS620 Thai                 | tis620_thai_ci      |      1 |
| euckr    | EUC-KR Korean               | euckr_korean_ci     |      2 |
| koi8u    | KOI8-U Ukrainian            | koi8u_general_ci    |      1 |
| gb2312   | GB2312 Simplified Chinese    | gb2312_chinese_ci   |      2 |
| greek    | ISO 8859-7 Greek            | greek_general_ci    |      1 |
| cp1250   | Windows Central European    | cp1250_general_ci   |      1 |
| gbk      | GBK Simplified Chinese       | gbk_chinese_ci      |      2 |
| latin5   | ISO 8859-9 Turkish          | latin5_turkish_ci   |      1 |
| armscii8 | ARMSCII-8 Armenian          | armscii8_general_ci |      1 |
| utf8     | UTF-8 Unicode               | utf8_general_ci     |      3 |
| ucs2     | UCS-2 Unicode               | ucs2_general_ci     |      2 |
| cp866    | DOS Russian                 | cp866_general_ci    |      1 |
```

```
| keybcs2  | DOS Kamenicky Czech-Slovak | keybcs2_general_ci  |       1 |
| macce    | Mac Central European       | macce_general_ci    |       1 |
| macroman | Mac West European          | macroman_general_ci |       1 |
| cp852    | DOS Central European       | cp852_general_ci    |       1 |
| latin7   | ISO 8859-13 Baltic         | latin7_general_ci   |       1 |
| cp1251   | Windows Cyrillic           | cp1251_general_ci   |       1 |
| cp1256   | Windows Arabic             | cp1256_general_ci   |       1 |
| cp1257   | Windows Baltic             | cp1257_general_ci   |       1 |
| binary   | Binary pseudo charset      | binary              |       1 |
| geostd8  | GEOSTD8 Georgian           | geostd8_general_ci  |       1 |
| cp932    | SJIS for Windows Japanese  | cp932_japanese_ci   |       2 |
| eucjpms  | UJIS for Windows Japanese  | eucjpms_japanese_ci |       3 |
+----------+----------------------------+---------------------+---------+
36 rows in set (0.11 sec)
```

If the value in the fourth column, maxlen, is greater than 1, then the character set is a multibyte character set.

When I installed the MySQL server, the latin1 character set was automatically chosen as the default character set. However, you may choose to use a different character set for each character column in your database, and you can even store different character sets within the same table. To choose a character set other than the default when defining a column, simply name one of the supported character sets after the type definition, as in:

```
varchar(20) character set utf8
```

With MySQL, you may also set the default character set for your entire database:

```
create database foreign_sales character set utf8;
```

While this is as much information regarding character sets as I'm willing to discuss in an introductory book, there is a great deal more to the topic of internationalization than what is shown here. If you plan to deal with multiple or unfamiliar character sets, you may want to pick up a book such as Andy Deitsch and David Czarnecki's *Java Internationalization (http://oreilly.com/catalog/9780596000196/)* (O'Reilly) or Richard Gillam's *Unicode Demystified: A Practical Programmer's Guide to the Encoding Standard* (Addison-Wesley).

Text data

If you need to store data that might exceed the 64 KB limit for varchar columns, you will need to use one of the text types.

Table 2-2 shows the available text types and their maximum sizes.

Table 2-2. MySQL text types

Text type	Maximum number of bytes
Tinytext	255
Text	65,535
Mediumtext	16,777,215
Longtext	4,294,967,295

When choosing to use one of the text types, you should be aware of the following:

- If the data being loaded into a text column exceeds the maximum size for that type, the data will be truncated.
- Trailing spaces will not be removed when data is loaded into the column.
- When using text columns for sorting or grouping, only the first 1,024 bytes are used, although this limit may be increased if necessary.
- The different text types are unique to MySQL. SQL Server has a single text type for large character data, whereas DB2 and Oracle use a data type called clob, for Character Large Object.
- Now that MySQL allows up to 65,535 bytes for varchar columns (it was limited to 255 bytes in version 4), there isn't any particular need to use the tinytext or text type.

If you are creating a column for free-form data entry, such as a notes column to hold data about customer interactions with your company's customer service department, then varchar will probably be adequate. If you are storing documents, however, you should choose either the mediumtext or longtext type.

> Oracle Database allows up to 2,000 bytes for char columns and 4,000 bytes for varchar2 columns. SQL Server can handle up to 8,000 bytes for both char and varchar data.

Numeric Data

Although it might seem reasonable to have a single numeric data type called "numeric," there are actually several different numeric data types that reflect the various ways in which numbers are used, as illustrated here:

A column indicating whether a customer order has been shipped
> This type of column, referred to as a *Boolean*, would contain a 0 to indicate false and a 1 to indicate true.

A system-generated primary key for a transaction table
> This data would generally start at 1 and increase in increments of one up to a potentially very large number.

An item number for a customer's electronic shopping basket
> The values for this type of column would be positive whole numbers between 1 and, at most, 200 (for shopaholics).

Positional data for a circuit board drill machine
> High-precision scientific or manufacturing data often requires accuracy to eight decimal points.

To handle these types of data (and more), MySQL has several different numeric data types. The most commonly used numeric types are those used to store whole numbers. When specifying one of these types, you may also specify that the data is *unsigned*, which tells the server that all data stored in the column will be greater than or equal to zero. Table 2-3 shows the five different data types used to store whole-number integers.

Table 2-3. MySQL integer types

Type	Signed range	Unsigned range
Tinyint	−128 to 127	0 to 255
Smallint	−32,768 to 32,767	0 to 65,535
Mediumint	−8,388,608 to 8,388,607	0 to 16,777,215
Int	−2,147,483,648 to 2,147,483,647	0 to 4,294,967,295
Bigint	−9,223,372,036,854,775,808 to 9,223,372,036,854,775,807	0 to 18,446,744,073,709,551,615

When you create a column using one of the integer types, MySQL will allocate an appropriate amount of space to store the data, which ranges from one byte for a `tinyint` to eight bytes for a `bigint`. Therefore, you should try to choose a type that will be large enough to hold the biggest number you can envision being stored in the column without needlessly wasting storage space.

For floating-point numbers (such as 3.1415927), you may choose from the numeric types shown in Table 2-4.

Table 2-4. MySQL floating-point types

Type	Numeric range
Float(p,s)	−3.402823466E+38 to −1.175494351E-38
	and 1.175494351E-38 to 3.402823466E+38
Double(p,s)	−1.7976931348623157E+308 to −2.2250738585072014E-308
	and 2.2250738585072014E-308 to 1.7976931348623157E+308

When using a floating-point type, you can specify a *precision* (the total number of allowable digits both to the left and to the right of the decimal point) and a *scale* (the number of allowable digits to the right of the decimal point), but they are not required. These values are represented in Table 2-4 as p and s. If you specify a precision and scale for your floating-point column, remember that the data stored in the column will be rounded if the number of digits exceeds the scale and/or precision of the column. For example, a column defined as float(4,2) will store a total of four digits, two to the left of the decimal and two to the right of the decimal. Therefore, such a column would handle the numbers 27.44 and 8.19 just fine, but the number 17.8675 would be rounded to 17.87, and attempting to store the number 178.375 in your float(4,2) column would generate an error.

Like the integer types, floating-point columns can be defined as unsigned, but this designation only prevents negative numbers from being stored in the column rather than altering the range of data that may be stored in the column.

Temporal Data

Along with strings and numbers, you will almost certainly be working with information about dates and/or times. This type of data is referred to as *temporal*, and some examples of temporal data in a database include:

- The future date that a particular event is expected to happen, such as shipping a customer's order
- The date that a customer's order was shipped
- The date and time that a user modified a particular row in a table
- An employee's birth date
- The year corresponding to a row in a yearly_sales fact table in a data warehouse
- The elapsed time needed to complete a wiring harness on an automobile assembly line

MySQL includes data types to handle all of these situations. Table 2-5 shows the temporal data types supported by MySQL.

Table 2-5. MySQL temporal types

Type	Default format	Allowable values
Date	YYYY-MM-DD	1000-01-01 to 9999-12-31
Datetime	YYYY-MM-DD HH:MI:SS	1000-01-01 00:00:00 to 9999-12-31 23:59:59
Timestamp	YYYY-MM-DD HH:MI:SS	1970-01-01 00:00:00 to 2037-12-31 23:59:59
Year	YYYY	1901 to 2155
Time	HHH:MI:SS	-838:59:59 to 838:59:59

While database servers store temporal data in various ways, the purpose of a format string (second column of Table 2-5) is to show how the data will be represented when retrieved, along with how a date string should be constructed when inserting or updating a temporal column. Thus, if you wanted to insert the date March 23, 2005 into a date column using the default format YYYY-MM-DD, you would use the string

'2005-03-23'. Chapter 7 fully explores how temporal data is constructed and displayed.

 Each database server allows a different range of dates for temporal columns. Oracle Database accepts dates ranging from 4712 BC to 9999 AD, while SQL Server only handles dates ranging from 1753 AD to 9999 AD (unless you are using SQL Server 2008's new `datetime2` data type, which allows for dates ranging from 1 AD to 9999 AD). MySQL falls in between Oracle and SQL Server and can store dates from 1000 AD to 9999 AD. Although this might not make any difference for most systems that track current and future events, it is important to keep in mind if you are storing historical dates.

Table 2-6 describes the various components of the date formats shown in Table 2-5.

Table 2-6. Date format components

Component	Definition	Range
YYYY	Year, including century	1000 to 9999
MM	Month	01 (January) to 12 (December)
DD	Day	01 to 31
HH	Hour	00 to 23
HHH	Hours (elapsed)	−838 to 838
MI	Minute	00 to 59
SS	Second	00 to 59

Here's how the various temporal types would be used to implement the examples shown earlier:

- Columns to hold the expected future shipping date of a customer order and an employee's birth date would use the `date` type, since it is unnecessary to know at what time a person was born and unrealistic to schedule a future shipment down to the second.

- A column to hold information about when a customer order was actually shipped would use the `datetime` type, since it is important to track not only the date that the shipment occurred but the time as well.

- A column that tracks when a user last modified a particular row in a table would use the `timestamp` type. The `timestamp` type holds the same information as the `datetime` type (year, month, day, hour, minute, second), but a `timestamp` column will automatically be populated with the current date/time by the MySQL server when a row is added to a table or when a row is later modified.

- A column holding just year data would use the `year` type.

- Columns that hold data regarding the length of time needed to complete a task would use the `time` type. For this type of data, it would be unnecessary and confusing to store a date component, since you are interested only in the number of hours/minutes/seconds needed to complete the task. This information could be derived using two `datetime` columns (one for the task start date/time and the other for the task completion date/time) and subtracting one from the other, but it is simpler to use a single `time` column.

Chapter 7 explores how to work with each of these temporal data types.

Table Creation

Now that you have a firm grasp on what data types may be stored in a MySQL database, it's time to see how to use these types in table definitions. Let's start by defining a table to hold information about a person.

Step 1: Design

A good way to start designing a table is to do a bit of brainstorming to see what kind of information would be helpful to include. Here's what I came up with after thinking for a short time about the types of information that describe a person:

- Name
- Gender
- Birth date
- Address
- Favorite foods

This is certainly not an exhaustive list, but it's good enough for now. The next step is to assign column names and data types. Table 2-7 shows my initial attempt.

Table 2-7. Person table, first pass

Column	Type	Allowable values
Name	Varchar(40)	
Gender	Char(1)	M, F
Birth_date	Date	
Address	Varchar(100)	
Favorite_foods	Varchar(200)	

The `name`, `address`, and `favorite_foods` columns are of type `varchar` and allow for free-form data entry. The `gender` column allows a single character which should equal only `M` or `F`. The `birth_date` column is of type `date`, since a time component is not needed.

Step 2: Refinement

In Chapter 1, you were introduced to the concept of *normalization*, which is the process of ensuring that there are no duplicate (other than foreign keys) or compound columns in your database design. In looking at the columns in the person table a second time, the following issues arise:

- The name column is actually a compound object consisting of a first name and a last name.
- Since multiple people can have the same name, gender, birth date, and so forth, there are no columns in the person table that guarantee uniqueness.
- The address column is also a compound object consisting of street, city, state/province, country, and postal code.
- The favorite_foods column is a list containing 0, 1, or more independent items. It would be best to create a separate table for this data that includes a foreign key to the person table so that you know to which person a particular food may be attributed.

After taking these issues into consideration, Table 2-8 gives a normalized version of the person table.

Table 2-8. Person table, second pass

Column	Type	Allowable values
Person_id	Smallint (unsigned)	
First_name	Varchar(20)	
Last_name	Varchar(20)	
Gender	Char(1)	M, F
Birth_date	Date	
Street	Varchar(30)	
City	Varchar(20)	
State	Varchar(20)	
Country	Varchar(20)	
Postal_code	Varchar(20)	

Now that the person table has a primary key (person_id) to guarantee uniqueness, the next step is to build a favorite_food table that includes a foreign key to the person table. Table 2-9 shows the result.

Table 2-9. Favorite_food table

Column	Type
Person_id	Smallint (unsigned)
Food	Varchar(20)

The person_id and food columns comprise the primary key of the favorite_food table, and the person_id column is also a foreign key to the person table.

How Much Is Enough?

Moving the favorite_foods column out of the person table was definitely a good idea, but are we done yet? What happens, for example, if one person lists "pasta" as a favorite food while another person lists "spaghetti"? Are they the same thing? In order to prevent this problem, you might decide that you want people to choose their favorite foods from a list of options, in which case you should create a food table with food_id and food_name columns, and then change the favorite_food table to contain a foreign key to the food table. While this design would be fully normalized, you might decide that you simply want to store the values that the user has entered, in which case you may leave the table as is.

Step 3: Building SQL Schema Statements

Now that the design is complete for the two tables holding information about people and their favorite foods, the next step is to generate SQL statements to create the tables in the database. Here is the statement to create the person table:

```
CREATE TABLE person
 (person_id SMALLINT UNSIGNED,
  fname VARCHAR(20),
  lname VARCHAR(20),
  gender CHAR(1),
  birth_date DATE,
  street VARCHAR(30),
  city VARCHAR(20),
  state VARCHAR(20),
  country VARCHAR(20),
  postal_code VARCHAR(20),
  CONSTRAINT pk_person PRIMARY KEY (person_id)
 );
```

Everything in this statement should be fairly self-explanatory except for the last item; when you define your table, you need to tell the database server what column or columns will serve as the primary key for the table. You do this by creating a *constraint* on the table. You can add several types of constraints to a table definition. This constraint is a *primary key constraint*. It is created on the person_id column and given the name pk_person.

While on the topic of constraints, there is another type of constraint that would be useful for the person table. In Table 2-7, I added a third column to show the allowable values for certain columns (such as 'M' and 'F' for the gender column). Another type of constraint called a *check constraint* constrains the allowable values for a particular column. MySQL allows a check constraint to be attached to a column definition, as in the following:

```
gender CHAR(1) CHECK (gender IN ('M','F')),
```

While check constraints operate as expected on most database servers, the MySQL server allows check constraints to be defined but does not enforce them. However, MySQL does provide another character data type called enum that merges the check constraint into the data type definition. Here's what it would look like for the gender column definition:

```
gender ENUM('M','F'),
```

Here's how the person table definition looks with an enum data type for the gender column:

```
CREATE TABLE person
 (person_id SMALLINT UNSIGNED,
  fname VARCHAR(20),
  lname VARCHAR(20),
  gender ENUM('M','F'),
  birth_date DATE,
  street VARCHAR(30),
  city VARCHAR(20),
  state VARCHAR(20),
  country VARCHAR(20),
  postal_code VARCHAR(20),
  CONSTRAINT pk_person PRIMARY KEY (person_id)
 );
```

Later in this chapter, you will see what happens if you try to add data to a column that violates its check constraint (or, in the case of MySQL, its enumeration values).

You are now ready to run the create table statement using the mysql command-line tool. Here's what it looks like:

```
mysql> CREATE TABLE person
    -> (person_id SMALLINT UNSIGNED,
    ->  fname VARCHAR(20),
    ->  lname VARCHAR(20),
    ->  gender ENUM('M','F'),
    ->  birth_date DATE,
    ->  street VARCHAR(30),
    ->  city VARCHAR(20),
    ->  state VARCHAR(20),
    ->  country VARCHAR(20),
    ->  postal_code VARCHAR(20),
    ->  CONSTRAINT pk_person PRIMARY KEY (person_id)
    -> );
Query OK, 0 rows affected (0.27 sec)
```

After processing the `create table` statement, the MySQL server returns the message "Query OK, 0 rows affected," which tells me that the statement had no syntax errors. If you want to make sure that the `person` table does, in fact, exist, you can use the `describe` command (or `desc` for short) to look at the table definition:

```
mysql> DESC person;
+-------------+----------------------+------+-----+---------+-------+
| Field       | Type                 | Null | Key | Default | Extra |
+-------------+----------------------+------+-----+---------+-------+
| person_id   | smallint(5) unsigned |      | PRI | 0       |       |
| fname       | varchar(20)          | YES  |     | NULL    |       |
| lname       | varchar(20)          | YES  |     | NULL    |       |
| gender      | enum('M','F')        | YES  |     | NULL    |       |
| birth_date  | date                 | YES  |     | NULL    |       |
| street      | varchar(30)          | YES  |     | NULL    |       |
| city        | varchar(20)          | YES  |     | NULL    |       |
| state       | varchar(20)          | YES  |     | NULL    |       |
| country     | varchar(20)          | YES  |     | NULL    |       |
| postal_code | varchar(20)          | YES  |     | NULL    |       |
+-------------+----------------------+------+-----+---------+-------+
10 rows in set (0.06 sec)
```

Columns 1 and 2 of the `describe` output are self-explanatory. Column 3 shows whether a particular column can be omitted when data is inserted into the table. I purposefully left this topic out of the discussion for now (see the sidebar "What Is Null?" on page 29 for a short discourse), but we explore it fully in Chapter 4. The fourth column shows whether a column takes part in any keys (primary or foreign); in this case, the `person_id` column is marked as the primary key. Column 5 shows whether a particular column will be populated with a default value if you omit the column when inserting data into the table. The `person_id` column shows a default value of 0, although this would work only once, since each row in the `person` table must contain a unique value for this column (since it is the primary key). The sixth column (called "Extra") shows any other pertinent information that might apply to a column.

What Is Null?

In some cases, it is not possible or applicable to provide a value for a particular column in your table. For example, when adding data about a new customer order, the `ship_date` column cannot yet be determined. In this case, the column is said to be *null* (note that I do not say that it *equals* null), which indicates the absence of a value. Null is used for various cases where a value cannot be supplied, such as:

- Not applicable
- Unknown
- Empty set

When designing a table, you may specify which columns are allowed to be null (the default), and which columns are not allowed to be null (designated by adding the keywords `not null` after the type definition).

Now that you've created the person table, your next step is to create the favorite_food table:

```
mysql> CREATE TABLE favorite_food
    -> (person_id SMALLINT UNSIGNED,
    ->  food VARCHAR(20),
    ->  CONSTRAINT pk_favorite_food PRIMARY KEY (person_id, food),
    ->  CONSTRAINT fk_fav_food_person_id FOREIGN KEY (person_id)
    ->    REFERENCES person (person_id)
    -> );
Query OK, 0 rows affected (0.10 sec)
```

This should look very similar to the create table statement for the person table, with the following exceptions:

- Since a person can have more than one favorite food (which is the reason this table was created in the first place), it takes more than just the person_id column to guarantee uniqueness in the table. This table, therefore, has a two-column primary key: person_id and food.

- The favorite_food table contains another type of constraint called a *foreign key constraint*. This constrains the values of the person_id column in the favorite_food table to include *only* values found in the person table. With this constraint in place, I will not be able to add a row to the favorite_food table indicating that person_id 27 likes pizza if there isn't already a row in the person table having a person_id of 27.

 If you forget to create the foreign key constraint when you first create the table, you can add it later via the alter table statement.

Describe shows the following after executing the create table statement:

```
mysql> DESC favorite_food;
+-------------+--------------------+------+-----+---------+-------+
| Field       | Type               | Null | Key | Default | Extra |
+-------------+--------------------+------+-----+---------+-------+
| person_id   | smallint(5) unsigned |    | PRI | 0       |       |
| food        | varchar(20)        |      | PRI |         |       |
+-------------+--------------------+------+-----+---------+-------+
```

Now that the tables are in place, the next logical step is to add some data.

Populating and Modifying Tables

With the person and favorite_food tables in place, you can now begin to explore the four SQL data statements: insert, update, delete, and select.

Inserting Data

Since there is not yet any data in the `person` and `favorite_food` tables, the first of the four SQL data statements to be explored will be the `insert` statement. There are three main components to an `insert` statement:

- The name of the table into which to add the data
- The names of the columns in the table to be populated
- The values with which to populate the columns

You are not required to provide data for every column in the table (unless all the columns in the table have been defined as `not null`). In some cases, those columns that are not included in the initial `insert` statement will be given a value later via an `update` statement. In other cases, a column may never receive a value for a particular row of data (such as a customer order that is canceled before being shipped, thus rendering the `ship_date` column inapplicable).

Generating numeric key data

Before inserting data into the `person` table, it would be useful to discuss how values are generated for numeric primary keys. Other than picking a number out of thin air, you have a couple of options:

- Look at the largest value currently in the table and add one.
- Let the database server provide the value for you.

Although the first option may seem valid, it proves problematic in a multiuser environment, since two users might look at the table at the same time and generate the same value for the primary key. Instead, all database servers on the market today provide a safe, robust method for generating numeric keys. In some servers, such as the Oracle Database, a separate schema object is used (called a *sequence*); in the case of MySQL, however, you simply need to turn on the *auto-increment* feature for your primary key column. Normally, you would do this at table creation, but doing it now provides the opportunity to learn another SQL schema statement, `alter table`, which is used to modify the definition of an existing table:

```
ALTER TABLE person MODIFY person_id SMALLINT UNSIGNED AUTO_INCREMENT;
```

This statement essentially redefines the `person_id` column in the `person` table. If you describe the table, you will now see the auto-increment feature listed under the "Extra" column for `person_id`:

```
mysql> DESC person;
+-------------+-----------------------+------+-----+---------+----------------+
| Field       | Type                  | Null | Key | Default | Extra          |
+-------------+-----------------------+------+-----+---------+----------------+
| person_id   | smallint(5) unsigned  |      | PRI | NULL    | auto_increment |
| .           |                       |      |     |         |                |
| .           |                       |      |     |         |                |
| .           |                       |      |     |         |                |
```

When you insert data into the person table, simply provide a null value for the per
son_id column, and MySQL will populate the column with the next available number
(by default, MySQL starts at 1 for auto-increment columns).

The insert statement

Now that all the pieces are in place, it's time to add some data. The following statement
creates a row in the person table for William Turner:

```
mysql> INSERT INTO person
    -> (person_id, fname, lname, gender, birth_date)
    -> VALUES (null, 'William','Turner', 'M', '1972-05-27');
Query OK, 1 row affected (0.01 sec)
```

The feedback ("Query OK, 1 row affected") tells you that your statement syntax was
proper, and that one row was added to the database (since it was an insert statement).
You can look at the data just added to the table by issuing a select statement:

```
mysql> SELECT person_id, fname, lname, birth_date
    -> FROM person;
+-----------+---------+--------+------------+
| person_id | fname   | lname  | birth_date |
+-----------+---------+--------+------------+
|         1 | William | Turner | 1972-05-27 |
+-----------+---------+--------+------------+
1 row in set (0.06 sec)
```

As you can see, the MySQL server generated a value of 1 for the primary key. Since
there is only a single row in the person table, I neglected to specify which row I am
interested in and simply retrieved all the rows in the table. If there were more than one
row in the table, however, I could add a where clause to specify that I want to retrieve
data only for the row having a value of 1 for the person_id column:

```
mysql> SELECT person_id, fname, lname, birth_date
    -> FROM person
    -> WHERE person_id = 1;
+-----------+---------+--------+------------+
| person_id | fname   | lname  | birth_date |
+-----------+---------+--------+------------+
|         1 | William | Turner | 1972-05-27 |
+-----------+---------+--------+------------+
1 row in set (0.00 sec)
```

While this query specifies a particular primary key value, you can use any column in the table to search for rows, as shown by the following query, which finds all rows with a value of 'Turner' for the lname column:

```
mysql> SELECT person_id, fname, lname, birth_date
    -> FROM person
    -> WHERE lname = 'Turner';
+-----------+---------+--------+------------+
| person_id | fname   | lname  | birth_date |
+-----------+---------+--------+------------+
|         1 | William | Turner | 1972-05-27 |
+-----------+---------+--------+------------+
1 row in set (0.00 sec)
```

Before moving on, a couple of things about the earlier insert statement are worth mentioning:

- Values were not provided for any of the address columns. This is fine, since nulls are allowed for those columns.
- The value provided for the birth_date column was a string. As long as you match the required format shown in Table 2-5, MySQL will convert the string to a date for you.
- The column names and the values provided must correspond in number and type. If you name seven columns and provide only six values, or if you provide values that cannot be converted to the appropriate data type for the corresponding column, you will receive an error.

William has also provided information about his favorite three foods, so here are three insert statements to store his food preferences:

```
mysql> INSERT INTO favorite_food (person_id, food)
    -> VALUES (1, 'pizza');
Query OK, 1 row affected (0.01 sec)
mysql> INSERT INTO favorite_food (person_id, food)
    -> VALUES (1, 'cookies');
Query OK, 1 row affected (0.00 sec)
mysql> INSERT INTO favorite_food (person_id, food)
    -> VALUES (1, 'nachos');
Query OK, 1 row affected (0.01 sec)
```

Here's a query that retrieves William's favorite foods in alphabetical order using an order by clause:

```
mysql> SELECT food
    -> FROM favorite_food
    -> WHERE person_id = 1
    -> ORDER BY food;
+---------+
| food    |
+---------+
| cookies |
| nachos  |
| pizza   |
```

```
      +---------+
      3 rows in set (0.02 sec)
```

The order by clause tells the server how to sort the data returned by the query. Without the order by clause, there is no guarantee that the data in the table will be retrieved in any particular order.

So that William doesn't get lonely, you can execute another insert statement to add Susan Smith to the person table:

```
mysql> INSERT INTO person
    ->   (person_id, fname, lname, gender, birth_date,
    ->    street, city, state, country, postal_code)
    -> VALUES (null, 'Susan','Smith', 'F', '1975-11-02',
    ->    '23 Maple St.', 'Arlington', 'VA', 'USA', '20220');
Query OK, 1 row affected (0.01 sec)
```

Since Susan was kind enough to provide her address, we included five more columns than when William's data was inserted. If you query the table again, you will see that Susan's row has been assigned the value 2 for its primary key value:

```
mysql> SELECT person_id, fname, lname, birth_date
    -> FROM person;
+-----------+---------+--------+------------+
| person_id | fname   | lname  | birth_date |
+-----------+---------+--------+------------+
|         1 | William | Turner | 1972-05-27 |
|         2 | Susan   | Smith  | 1975-11-02 |
+-----------+---------+--------+------------+
2 rows in set (0.00 sec)
```

Can I Get That in XML?

If you will be working with XML data, you will be happy to know that most database servers provide a simple way to generate XML output from a query. With MySQL, for example, you can use the --xml option when invoking the mysql tool, and all your output will automatically be formatted using XML. Here's what the favorite-food data looks like as an XML document:

```
C:\database> mysql -u lrngsql -p --xml bank
Enter password: xxxxxx
Welcome to the MySQL Monitor...

Mysql> SELECT * FROM favorite_food;
<?xml version="1.0"?>

<resultset statement="select * from favorite_food"
xmlns:xsi="http://www.w3.org/2001/XMLSchema-instance">
  <row>
        <field name="person_id">1</field>
        <field name="food">cookies</field>
  </row>
  <row>
        <field name="person_id">1</field>
        <field name="food">nachos</field>
  </row>
```

```
        <row>
              <field name="person_id">1</field>
              <field name="food">pizza</field>
        </row>
</resultset>
3 rows in set (0.00 sec)
```

With SQL Server, you don't need to configure your command-line tool; you just need to add the `for xml` clause to the end of your query, as in:

```
SELECT * FROM favorite_food
FOR XML AUTO, ELEMENTS
```

Updating Data

When the data for William Turner was initially added to the table, data for the various address columns was omitted from the `insert` statement. The next statement shows how these columns can be populated via an `update` statement:

```
mysql> UPDATE person
    -> SET street = '1225 Tremont St.',
    ->    city = 'Boston',
    ->    state = 'MA',
    ->    country = 'USA',
    ->    postal_code = '02138'
    -> WHERE person_id = 1;
Query OK, 1 row affected (0.04 sec)
Rows matched: 1  Changed: 1  Warnings: 0
```

The server responded with a two-line message: the "Rows matched: 1" item tells you that the condition in the `where` clause matched a single row in the table, and the "Changed: 1" item tells you that a single row in the table has been modified. Since the `where` clause specifies the primary key of William's row, this is exactly what you would expect to have happen.

Depending on the conditions in your `where` clause, it is also possible to modify more than one row using a single statement. Consider, for example, what would happen if your `where` clause looked as follows:

```
WHERE person_id < 10
```

Since both William and Susan have a `person_id` value less than 10, both of their rows would be modified. If you leave off the `where` clause altogether, your `update` statement will modify every row in the table.

Deleting Data

It seems that William and Susan aren't getting along very well together, so one of them has got to go. Since William was there first, Susan will get the boot courtesy of the `delete` statement:

```
mysql> DELETE FROM person
    -> WHERE person_id = 2;
Query OK, 1 row affected (0.01 sec)
```

Again, the primary key is being used to isolate the row of interest, so a single row is deleted from the table. Similar to the `update` statement, more than one row can be deleted depending on the conditions in your `where` clause, and all rows will be deleted if the `where` clause is omitted.

When Good Statements Go Bad

So far, all of the SQL data statements shown in this chapter have been well formed and have played by the rules. Based on the table definitions for the `person` and `favorite_food` tables, however, there are lots of ways that you can run afoul when inserting or modifying data. This section shows you some of the common mistakes that you might come across and how the MySQL server will respond.

Nonunique Primary Key

Because the table definitions include the creation of primary key constraints, MySQL will make sure that duplicate key values are not inserted into the tables. The next statement attempts to bypass the auto-increment feature of the `person_id` column and create another row in the `person` table with a `person_id` of 1:

```
mysql> INSERT INTO person
    -> (person_id, fname, lname, gender, birth_date)
    -> VALUES (1, 'Charles','Fulton', 'M', '1968-01-15');
ERROR 1062 (23000): Duplicate entry '1' for key 'PRIMARY'
```

There is nothing stopping you (with the current schema objects, at least) from creating two rows with identical names, addresses, birth dates, and so on, as long as they have different values for the `person_id` column.

Nonexistent Foreign Key

The table definition for the `favorite_food` table includes the creation of a foreign key constraint on the `person_id` column. This constraint ensures that all values of `person_id` entered into the `favorite_food` table exist in the `person` table. Here's what would happen if you tried to create a row that violates this constraint:

```
mysql> INSERT INTO favorite_food (person_id, food)
    -> VALUES (999, 'lasagna');
ERROR 1452 (23000): Cannot add or update a child row: a foreign key constraint
fails ('bank'.'favorite_food', CONSTRAINT 'fk_fav_food_person_id' FOREIGN KEY
('person_id') REFERENCES 'person' ('person_id'))
```

In this case, the `favorite_food` table is considered the *child* and the `person` table is considered the *parent*, since the `favorite_food` table is dependent on the `person` table

---

for some of its data. If you plan to enter data into both tables, you will need to create a row in `parent` before you can enter data into `favorite_food`.

 Foreign key constraints are enforced only if your tables are created using the InnoDB storage engine. We discuss MySQL's storage engines in Chapter 12.

Column Value Violations

The `gender` column in the `person` table is restricted to the values `'M'` for male and `'F'` for female. If you mistakenly attempt to set the value of the column to any other value, you will receive the following response:

```
mysql> UPDATE person
    -> SET gender = 'Z'
    -> WHERE person_id = 1;
ERROR 1265 (01000): Data truncated for column 'gender' at row 1
```

The error message is a bit confusing, but it gives you the general idea that the server is unhappy about the value provided for the `gender` column.

Invalid Date Conversions

If you construct a string with which to populate a `date` column, and that string does not match the expected format, you will receive another error. Here's an example that uses a date format that does not match the default date format of "YYYY-MM-DD":

```
mysql> UPDATE person
    -> SET birth_date = 'DEC-21-1980'
    -> WHERE person_id = 1;
ERROR 1292 (22007): Incorrect date value: 'DEC-21-1980' for column 'birth_date'
at row 1
```

In general, it is always a good idea to explicitly specify the format string rather than relying on the default format. Here's another version of the statement that uses the `str_to_date` function to specify which format string to use:

```
mysql> UPDATE person
    -> SET birth_date = str_to_date('DEC-21-1980' , '%b-%d-%Y')
    -> WHERE person_id = 1;
Query OK, 1 row affected (0.12 sec)
Rows matched: 1  Changed: 1  Warnings: 0
```

Not only is the database server happy, but William is happy as well (we just made him eight years younger, without the need for expensive cosmetic surgery!).

Earlier in the chapter, when I discussed the various temporal data types, I showed date-formatting strings such as "YYYY-MM-DD". While many database servers use this style of formatting, MySQL uses %Y to indicate a four-character year. Here are a few more formatters that you might need when converting strings to datetimes in MySQL:

```
%a The short weekday name, such as Sun, Mon, ...
%b The short month name, such as Jan, Feb, ...
%c The numeric month (0..12)
%d The numeric day of the month (00..31)
%f The number of microseconds (000000..999999)
%H The hour of the day, in 24-hour format (00..23)
%h The hour of the day, in 12-hour format (01..12)
%i The minutes within the hour (00..59)
%j The day of year (001..366)
%M The full month name (January..December)
%m The numeric month
%p AM or PM
%s The number of seconds (00..59)
%W The full weekday name (Sunday..Saturday)
%w The numeric day of the week (0=Sunday..6=Saturday)
%Y The four-digit year
```

The Bank Schema

For the remainder of the book, you use a group of tables that model a community bank. Some of the tables include Employee, Branch, Account, Customer, Product, and Transaction. The entire schema and example data should have been created when you followed the final steps at the beginning of the chapter for loading the MySQL server and generating the sample data. To see a diagram of the tables and their columns and relationships, see Appendix A.

Table 2-10 shows all the tables used in the bank schema along with short definitions.

Table 2-10. Bank schema definitions

Table name	Definition
Account	A particular product opened for a particular customer
Branch	A location at which banking transactions are conducted
Business	A corporate customer (subtype of the Customer table)
Customer	A person or corporation known to the bank
Department	A group of bank employees implementing a particular banking function
Employee	A person working for the bank
Individual	A noncorporate customer (subtype of the Customer table)
Officer	A person allowed to transact business for a corporate customer
Product	A banking service offered to customers
Product_type	A group of products having a similar function
Transaction	A change made to an account balance

Feel free to experiment with the tables as much as you want, including adding your own tables to expand the bank's business functions. You can always drop the database and re-create it from the downloaded file if you want to make sure your sample data is intact.

If you want to see the tables available in your database, you can use the show tables command, as in:

```
mysql> SHOW TABLES;
+-----------------+
| Tables_in_bank |
+-----------------+
| account         |
| branch          |
| business        |
| customer        |
| department      |
| employee        |
| favorite_food   |
| individual      |
| officer         |
| person          |
| product         |
| product_type    |
| transaction     |
+-----------------+
13 rows in set (0.10 sec)
```

Along with the 11 tables in the bank schema, the table listing also includes the two tables created in this chapter: person and favorite_food. These tables will not be used in later chapters, so feel free to drop them by issuing the following commands:

```
mysql> DROP TABLE favorite_food;
Query OK, 0 rows affected (0.56 sec)
mysql> DROP TABLE person;
Query OK, 0 rows affected (0.05 sec)
```

If you want to look at the columns in a table, you can use the describe command. Here's an example of the describe output for the customer table:

```
mysql> DESC customer;
+--------------+------------------+------+-----+---------+----------------+
| Field        | Type             | Null | Key | Default | Extra          |
+--------------+------------------+------+-----+---------+----------------+
| cust_id      | int(10) unsigned | NO   | PRI | NULL    | auto_increment |
| fed_id       | varchar(12)      | NO   |     | NULL    |                |
| cust_type_cd | enum('I','B')    | NO   |     | NULL    |                |
| address      | varchar(30)      | YES  |     | NULL    |                |
| city         | varchar(20)      | YES  |     | NULL    |                |
| state        | varchar(20)      | YES  |     | NULL    |                |
| postal_code  | varchar(10)      | YES  |     | NULL    |                |
+--------------+------------------+------+-----+---------+----------------+
7 rows in set (0.03 sec)
```

The more comfortable you are with the example database, the better you will understand the examples and, consequently, the concepts in the following chapters.

CHAPTER 3

Query Primer

So far, you have seen a few examples of database queries (a.k.a. `select` statements) sprinkled throughout the first two chapters. Now it's time to take a closer look at the different parts of the `select` statement and how they interact.

Query Mechanics

Before dissecting the `select` statement, it might be interesting to look at how queries are executed by the MySQL server (or, for that matter, any database server). If you are using the `mysql` command-line tool (which I assume you are), then you have already logged in to the MySQL server by providing your username and password (and possibly a hostname if the MySQL server is running on a different computer). Once the server has verified that your username and password are correct, a *database connection* is generated for you to use. This connection is held by the application that requested it (which, in this case, is the `mysql` tool) until the application releases the connection (i.e., as a result of your typing `quit`) or the server closes the connection (i.e., when the server is shut down). Each connection to the MySQL server is assigned an identifier, which is shown to you when you first log in:

```
Welcome to the MySQL monitor.  Commands end with ; or \g.
Your MySQL connection id is 11
Server version: 6.0.3-alpha-community MySQL Community Server (GPL)

Type 'help;' or '\h' for help. Type '\c' to clear the buffer.
```

In this case, my connection ID is 11. This information might be useful to your database administrator if something goes awry, such as a malformed query that runs for hours, so you might want to jot it down.

Once the server has verified your username and password and issued you a connection, you are ready to execute queries (along with other SQL statements). Each time a query is sent to the server, the server checks the following things prior to statement execution:

- Do you have permission to execute the statement?
- Do you have permission to access the desired data?
- Is your statement syntax correct?

If your statement passes these three tests, then your query is handed to the *query optimizer*, whose job it is to determine the most efficient way to execute your query. The optimizer will look at such things as the order in which to join the tables named in your from clause and what indexes are available, and then picks an *execution plan*, which the server uses to execute your query.

 Understanding and influencing how your database server chooses execution plans is a fascinating topic that many of you will wish to explore. For those readers using MySQL, you might consider reading Baron Schwartz et al.'s *High Performance MySQL (http://oreilly.com/catalog/9780596101718/)* (O'Reilly). Among other things, you will learn how to generate indexes, analyze execution plans, influence the optimizer via query hints, and tune your server's startup parameters. If you are using Oracle Database or SQL Server, dozens of tuning books are available.

Once the server has finished executing your query, the *result set* is returned to the calling application (which is, once again, the mysql tool). As I mentioned in Chapter 1, a result set is just another table containing rows and columns. If your query fails to yield any results, the mysql tool will show you the message found at the end of the following example:

```
mysql> SELECT emp_id, fname, lname
    -> FROM employee
    -> WHERE lname = 'Bkadfl';
Empty set(0.00 sec)
```

If the query returns one or more rows, the mysql tool will format the results by adding column headers and by constructing boxes around the columns using the -, |, and + symbols, as shown in the next example:

```
mysql> SELECT fname, lname
    -> FROM employee;
+----------+-----------+
| fname    | lname     |
+----------+-----------+
| Michael  | Smith     |
| Susan    | Barker    |
| Robert   | Tyler     |
| Susan    | Hawthorne |
| John     | Gooding   |
| Helen    | Fleming   |
| Chris    | Tucker    |
| Sarah    | Parker    |
| Jane     | Grossman  |
```

```
| Paula    | Roberts  |
| Thomas   | Ziegler  |
| Samantha | Jameson  |
| John     | Blake    |
| Cindy    | Mason    |
| Frank    | Portman  |
| Theresa  | Markham  |
| Beth     | Fowler   |
| Rick     | Tulman   |
+----------+----------+
18 rows in set (0.00 sec)
```

This query returns the first and last names of all the employees in the `employee` table. After the last row of data is displayed, the `mysql` tool displays a message telling you how many rows were returned, which, in this case, is 18.

Query Clauses

Several components or *clauses* make up the `select` statement. While only one of them is mandatory when using MySQL (the `select` clause), you will usually include at least two or three of the six available clauses. Table 3-1 shows the different clauses and their purposes.

Table 3-1. Query clauses

Clause name	Purpose
Select	Determines which columns to include in the query's result set
From	Identifies the tables from which to draw data and how the tables should be joined
Where	Filters out unwanted data
Group by	Used to group rows together by common column values
Having	Filters out unwanted groups
Order by	Sorts the rows of the final result set by one or more columns

All of the clauses shown in Table 3-1 are included in the ANSI specification; additionally, several other clauses are unique to MySQL that we explore in Appendix B. The following sections delve into the uses of the six major query clauses.

The select Clause

Even though the `select` clause is the first clause of a `select` statement, it is one of the last clauses that the database server evaluates. The reason for this is that before you can determine what to include in the final result set, you need to know all of the possible columns that *could* be included in the final result set. In order to fully understand the role of the `select` clause, therefore, you will need to understand a bit about the `from` clause. Here's a query to get started:

```
mysql> SELECT *
    -> FROM department;
+---------+----------------+
| dept_id | name           |
+---------+----------------+
|       1 | Operations     |
|       2 | Loans          |
|       3 | Administration |
+---------+----------------+
3 rows in set (0.04 sec)
```

In this query, the from clause lists a single table (department), and the select clause indicates that *all* columns (designated by *) in the department table should be included in the result set. This query could be described in English as follows:

Show me all the columns and all the rows in the department *table.*

In addition to specifying all the columns via the asterisk character, you can explicitly name the columns you are interested in, such as:

```
mysql> SELECT dept_id, name
    -> FROM department;
+---------+----------------+
| dept_id | name           |
+---------+----------------+
|       1 | Operations     |
|       2 | Loans          |
|       3 | Administration |
+---------+----------------+
3 rows in set (0.01 sec)
```

The results are identical to the first query, since all the columns in the department table (dept_id and name) are named in the select clause. You can choose to include only a subset of the columns in the department table as well:

```
mysql> SELECT name
    -> FROM department;
+----------------+
| name           |
+----------------+
| Operations     |
| Loans          |
| Administration |
+----------------+
3 rows in set (0.00 sec)
```

The job of the select clause, therefore, is the following:

The select *clause determines which of all possible columns should be included in the query's result set.*

If you were limited to including only columns from the table or tables named in the from clause, things would be rather dull. However, you can spice things up by including in your select clause such things as:

- Literals, such as numbers or strings
- Expressions, such as `transaction.amount * -1`
- Built-in function calls, such as `ROUND(transaction.amount, 2)`
- User-defined function calls

The next query demonstrates the use of a table column, a literal, an expression, and a built-in function call in a single query against the `employee` table:

```
mysql> SELECT emp_id,
    ->   'ACTIVE',
    ->   emp_id * 3.14159,
    ->   UPPER(lname)
    -> FROM employee;
+--------+--------+------------------+--------------+
| emp_id | ACTIVE | emp_id * 3.14159 | UPPER(lname) |
+--------+--------+------------------+--------------+
|      1 | ACTIVE |          3.14159 | SMITH        |
|      2 | ACTIVE |          6.28318 | BARKER       |
|      3 | ACTIVE |          9.42477 | TYLER        |
|      4 | ACTIVE |         12.56636 | HAWTHORNE    |
|      5 | ACTIVE |         15.70795 | GOODING      |
|      6 | ACTIVE |         18.84954 | FLEMING      |
|      7 | ACTIVE |         21.99113 | TUCKER       |
|      8 | ACTIVE |         25.13272 | PARKER       |
|      9 | ACTIVE |         28.27431 | GROSSMAN     |
|     10 | ACTIVE |         31.41590 | ROBERTS      |
|     11 | ACTIVE |         34.55749 | ZIEGLER      |
|     12 | ACTIVE |         37.69908 | JAMESON      |
|     13 | ACTIVE |         40.84067 | BLAKE        |
|     14 | ACTIVE |         43.98226 | MASON        |
|     15 | ACTIVE |         47.12385 | PORTMAN      |
|     16 | ACTIVE |         50.26544 | MARKHAM      |
|     17 | ACTIVE |         53.40703 | FOWLER       |
|     18 | ACTIVE |         56.54862 | TULMAN       |
+--------+--------+------------------+--------------+
18 rows in set (0.05 sec)
```

We cover expressions and built-in functions in detail later, but I wanted to give you a feel for what kinds of things can be included in the `select` clause. If you only need to execute a built-in function or evaluate a simple expression, you can skip the `from` clause entirely. Here's an example:

```
mysql> SELECT VERSION(),
    ->   USER(),
    ->   DATABASE();
+----------------------+-------------------+------------+
| version()            | user()            | database() |
+----------------------+-------------------+------------+
| 6.0.3-alpha-community | lrngsql@localhost | bank       |
+----------------------+-------------------+------------+
1 row in set (0.05 sec)
```

Since this query simply calls three built-in functions and doesn't retrieve data from any tables, there is no need for a `from` clause.

Column Aliases

Although the `mysql` tool will generate labels for the columns returned by your queries, you may want to assign your own labels. While you might want to assign a new label to a column from a table (if it is poorly or ambiguously named), you will almost certainly want to assign your own labels to those columns in your result set that are generated by expressions or built-in function calls. You can do so by adding a *column alias* after each element of your `select` clause. Here's the previous query against the `employee` table with column aliases applied to three of the columns:

```
mysql> SELECT emp_id,
    ->   'ACTIVE' status,
    ->   emp_id * 3.14159 empid_x_pi,
    ->   UPPER(lname) last_name_upper
    -> FROM employee;
+--------+--------+------------+-----------------+
| emp_id | status | empid_x_pi | last_name_upper |
+--------+--------+------------+-----------------+
|      1 | ACTIVE |    3.14159 | SMITH           |
|      2 | ACTIVE |    6.28318 | BARKER          |
|      3 | ACTIVE |    9.42477 | TYLER           |
|      4 | ACTIVE |   12.56636 | HAWTHORNE       |
|      5 | ACTIVE |   15.70795 | GOODING         |
|      6 | ACTIVE |   18.84954 | FLEMING         |
|      7 | ACTIVE |   21.99113 | TUCKER          |
|      8 | ACTIVE |   25.13272 | PARKER          |
|      9 | ACTIVE |   28.27431 | GROSSMAN        |
|     10 | ACTIVE |   31.41590 | ROBERTS         |
|     11 | ACTIVE |   34.55749 | ZIEGLER         |
|     12 | ACTIVE |   37.69908 | JAMESON         |
|     13 | ACTIVE |   40.84067 | BLAKE           |
|     14 | ACTIVE |   43.98226 | MASON           |
|     15 | ACTIVE |   47.12385 | PORTMAN         |
|     16 | ACTIVE |   50.26544 | MARKHAM         |
|     17 | ACTIVE |   53.40703 | FOWLER          |
|     18 | ACTIVE |   56.54862 | TULMAN          |
+--------+--------+------------+-----------------+
18 rows in set (0.00 sec)
```

If you look at the column headers, you can see that the second, third, and fourth columns now have reasonable names instead of simply being labeled with the function or expression that generated the column. If you look at the `select` clause, you can see how the column aliases `status`, `empid_x_pi`, and `last_name_upper` are added after the second, third, and fourth columns. I think you will agree that the output is easier to understand with column aliases in place, and it would be easier to work with programmatically if you were issuing the query from within Java or C# rather than interactively via the `mysql` tool. In order to make your column aliases stand out even more, you also have the option of using the **as** keyword before the alias name, as in:

```
mysql> SELECT emp_id,
    ->    'ACTIVE' AS status,
    ->    emp_id * 3.14159 AS empid_x_pi,
    ->    UPPER(lname) AS last_name_upper
    -> FROM employee;
```

Many people feel that including the optional **as** keyword improves readability, although I have chosen not to use it for the examples in this book.

Removing Duplicates

In some cases, a query might return duplicate rows of data. For example, if you were to retrieve the IDs of all customers that have accounts, you would see the following:

```
mysql> SELECT cust_id
    -> FROM account;
+---------+
| cust_id |
+---------+
|       1 |
|       1 |
|       1 |
|       2 |
|       2 |
|       3 |
|       3 |
|       4 |
|       4 |
|       4 |
|       5 |
|       6 |
|       6 |
|       7 |
|       8 |
|       8 |
|       9 |
|       9 |
|       9 |
|      10 |
|      10 |
|      11 |
|      12 |
|      13 |
+---------+
24 rows in set (0.00 sec)
```

Since some customers have more than one account, you will see the same customer ID once for each account owned by that customer. What you probably want in this case is the *distinct* set of customers that have accounts, instead of seeing the customer ID for each row in the `account` table. You can achieve this by adding the keyword `distinct` directly after the `select` keyword, as demonstrated by the following:

```
mysql> SELECT DISTINCT cust_id
    -> FROM account;
+---------+
| cust_id |
+---------+
|       1 |
|       2 |
|       3 |
|       4 |
|       5 |
|       6 |
|       7 |
|       8 |
|       9 |
|      10 |
|      11 |
|      12 |
|      13 |
+---------+
13 rows in set (0.01 sec)
```

The result set now contains 13 rows, one for each distinct customer, rather than 24 rows, one for each account.

If you do not want the server to remove duplicate data, or you are sure there will be no duplicates in your result set, you can specify the ALL keyword instead of specifying DISTINCT. However, the ALL keyword is the default and never needs to be explicitly named, so most programmers do not include ALL in their queries.

 Keep in mind that generating a distinct set of results requires the data to be sorted, which can be time-consuming for large result sets. Don't fall into the trap of using DISTINCT just to be sure there are no duplicates; instead, take the time to understand the data you are working with so that you will know whether duplicates are possible.

The from Clause

Thus far, you have seen queries whose `from` clauses contain a single table. Although most SQL books will define the `from` clause as simply a list of one or more tables, I would like to broaden the definition as follows:

The from *clause defines the tables used by a query, along with the means of linking the tables together*.

This definition is composed of two separate but related concepts, which we explore in the following sections.

Tables

When confronted with the term *table*, most people think of a set of related rows stored in a database. While this does describe one type of table, I would like to use the word in a more general way by removing any notion of how the data might be stored and concentrating on just the set of related rows. Three different types of tables meet this relaxed definition:

- Permanent tables (i.e., created using the `create table` statement)
- Temporary tables (i.e., rows returned by a subquery)
- Virtual tables (i.e., created using the `create view` statement)

Each of these table types may be included in a query's `from` clause. By now, you should be comfortable with including a permanent table in a `from` clause, so I briefly describe the other types of tables that can be referenced in a `from` clause.

Subquery-generated tables

A subquery is a query contained within another query. Subqueries are surrounded by parentheses and can be found in various parts of a `select` statement; within the `from` clause, however, a subquery serves the role of generating a temporary table that is visible from all other query clauses and can interact with other tables named in the `from` clause. Here's a simple example:

```
mysql> SELECT e.emp_id, e.fname, e.lname
    -> FROM (SELECT emp_id, fname, lname, start_date, title
    ->       FROM employee) e;
+--------+----------+-----------+
| emp_id | fname    | lname     |
+--------+----------+-----------+
|      1 | Michael  | Smith     |
|      2 | Susan    | Barker    |
|      3 | Robert   | Tyler     |
|      4 | Susan    | Hawthorne |
|      5 | John     | Gooding   |
|      6 | Helen    | Fleming   |
|      7 | Chris    | Tucker    |
|      8 | Sarah    | Parker    |
|      9 | Jane     | Grossman  |
|     10 | Paula    | Roberts   |
|     11 | Thomas   | Ziegler   |
|     12 | Samantha | Jameson   |
|     13 | John     | Blake     |
|     14 | Cindy    | Mason     |
|     15 | Frank    | Portman   |
|     16 | Theresa  | Markham   |
|     17 | Beth     | Fowler    |
|     18 | Rick     | Tulman    |
+--------+----------+-----------+
18 rows in set (0.00 sec)
```

In this example, a subquery against the employee table returns five columns, and the *containing query* references three of the five available columns. The subquery is referenced by the containing query via its alias, which, in this case, is e. This is a simplistic and not particularly useful example of a subquery in a from clause; you will find detailed coverage of subqueries in Chapter 9.

Views

A view is a query that is stored in the data dictionary. It looks and acts like a table, but there is no data associated with a view (this is why I call it a *virtual* table). When you issue a query against a view, your query is merged with the view definition to create a final query to be executed.

To demonstrate, here's a view definition that queries the employee table and includes a call to a built-in function:

```
mysql> CREATE VIEW employee_vw AS
    -> SELECT emp_id, fname, lname,
    ->   YEAR(start_date) start_year
    -> FROM employee;
Query OK, 0 rows affected (0.10 sec)
```

When the view is created, no additional data is generated or stored: the server simply tucks away the select statement for future use. Now that the view exists, you can issue queries against it, as in:

```
mysql> SELECT emp_id, start_year
    -> FROM employee_vw;
+--------+------------+
| emp_id | start_year |
+--------+------------+
|      1 |       2005 |
|      2 |       2006 |
|      3 |       2005 |
|      4 |       2006 |
|      5 |       2007 |
|      6 |       2008 |
|      7 |       2008 |
|      8 |       2006 |
|      9 |       2006 |
|     10 |       2006 |
|     11 |       2004 |
|     12 |       2007 |
|     13 |       2004 |
|     14 |       2006 |
|     15 |       2007 |
|     16 |       2005 |
|     17 |       2006 |
|     18 |       2006 |
+--------+------------+
18 rows in set (0.07 sec)
```

Views are created for various reasons, including to hide columns from users and to simplify complex database designs.

Table Links

The second deviation from the simple `from` clause definition is the mandate that if more than one table appears in the `from` clause, the conditions used to *link* the tables must be included as well. This is not a requirement of MySQL or any other database server, but it is the ANSI-approved method of joining multiple tables, and it is the most portable across the various database servers. We explore joining multiple tables in depth in Chapters 5 and 10, but here's a simple example in case I have piqued your curiosity:

```
mysql> SELECT employee.emp_id, employee.fname,
    ->    employee.lname, department.name dept_name
    -> FROM employee INNER JOIN department
    ->    ON employee.dept_id = department.dept_id;
+--------+----------+-----------+----------------+
| emp_id | fname    | lname     | dept_name      |
+--------+----------+-----------+----------------+
|      1 | Michael  | Smith     | Administration |
|      2 | Susan    | Barker    | Administration |
|      3 | Robert   | Tyler     | Administration |
|      4 | Susan    | Hawthorne | Operations     |
|      5 | John     | Gooding   | Loans          |
|      6 | Helen    | Fleming   | Operations     |
|      7 | Chris    | Tucker    | Operations     |
|      8 | Sarah    | Parker    | Operations     |
|      9 | Jane     | Grossman  | Operations     |
|     10 | Paula    | Roberts   | Operations     |
|     11 | Thomas   | Ziegler   | Operations     |
|     12 | Samantha | Jameson   | Operations     |
|     13 | John     | Blake     | Operations     |
|     14 | Cindy    | Mason     | Operations     |
|     15 | Frank    | Portman   | Operations     |
|     16 | Theresa  | Markham   | Operations     |
|     17 | Beth     | Fowler    | Operations     |
|     18 | Rick     | Tulman    | Operations     |
+--------+----------+-----------+----------------+
18 rows in set (0.05 sec)
```

The previous query displays data from both the `employee` table (`emp_id`, `fname`, `lname`) and the `department` table (`name`), so both tables are included in the `from` clause. The mechanism for linking the two tables (referred to as a *join*) is the employee's department affiliation stored in the `employee` table. Thus, the database server is instructed to use the value of the `dept_id` column in the `employee` table to look up the associated department name in the `department` table. Join conditions for two tables are found in the `on` subclause of the `from` clause; in this case, the join condition is `ON employee.dept_id = department.dept_id`. Again, please refer to Chapter 5 for a thorough discussion of joining multiple tables.

Defining Table Aliases

When multiple tables are joined in a single query, you need a way to identify which table you are referring to when you reference columns in the select, where, group by, having, and order by clauses. You have two choices when referencing a table outside the from clause:

- Use the entire table name, such as employee.emp_id.
- Assign each table an *alias* and use the alias throughout the query.

In the previous query, I chose to use the entire table name in the select and on clauses. Here's what the same query looks like using table aliases:

```
SELECT e.emp_id, e.fname, e.lname,
  d.name dept_name
FROM employee e INNER JOIN department d
  ON e.dept_id = d.dept_id;
```

If you look closely at the from clause, you will see that the employee table is assigned the alias e, and the department table is assigned the alias d. These aliases are then used in the on clause when defining the join condition as well as in the select clause when specifying the columns to include in the result set. I hope you will agree that using aliases makes for a more compact statement without causing confusion (as long as your choices for alias names are reasonable). Additionally, you may use the as keyword with your table aliases, similar to what was demonstrated earlier for column aliases:

```
SELECT e.emp_id, e.fname, e.lname,
  d.name dept_name
FROM employee AS e INNER JOIN department AS d
  ON e.dept_id = d.dept_id;
```

I have found that roughly half of the database developers I have worked with use the as keyword with their column and table aliases, and half do not.

The where Clause

The queries shown thus far in the chapter have selected every row from the employee, department, or account table (except for the demonstration of distinct earlier in the chapter). Most of the time, however, you will not wish to retrieve *every* row from a table but will want a way to filter out those rows that are not of interest. This is a job for the where clause.

The where clause is the mechanism for filtering out unwanted rows from your result set.

For example, perhaps you are interested in retrieving data from the employee table, but only for those employees who are employed as head tellers. The following query employs a where clause to retrieve *only* the four head tellers:

```
mysql> SELECT emp_id, fname, lname, start_date, title
    -> FROM employee
    -> WHERE title = 'Head Teller';
+--------+---------+---------+------------+-------------+
| emp_id | fname   | lname   | start_date | title       |
+--------+---------+---------+------------+-------------+
|      6 | Helen   | Fleming | 2008-03-17 | Head Teller |
|     10 | Paula   | Roberts | 2006-07-27 | Head Teller |
|     13 | John    | Blake   | 2004-05-11 | Head Teller |
|     16 | Theresa | Markham | 2005-03-15 | Head Teller |
+--------+---------+---------+------------+-------------+
4 rows in set (1.17 sec)
```

In this case the where clause filtered out 14 of the 18 employee rows. This where clause contains a single *filter condition*, but you can include as many conditions as required; individual conditions are separated using operators such as and, or, and not (see Chapter 4 for a complete discussion of the where clause and filter conditions). Here's an extension of the previous query that includes a second condition stating that only those employees with a start date later than January 1, 2006 should be included:

```
mysql> SELECT emp_id, fname, lname, start_date, title
    -> FROM employee
    -> WHERE title = 'Head Teller'
    ->   AND start_date > '2006-01-01';
+--------+-------+---------+------------+-------------+
| emp_id | fname | lname   | start_date | title       |
+--------+-------+---------+------------+-------------+
|      6 | Helen | Fleming | 2008-03-17 | Head Teller |
|     10 | Paula | Roberts | 2006-07-27 | Head Teller |
+--------+-------+---------+------------+-------------+
2 rows in set (0.01 sec)
```

The first condition (title = 'Head Teller') filtered out 14 of 18 employee rows, and the second condition (start_date > '2006-01-01') filtered out an additional 2 rows, leaving 2 rows in the final result set. Let's see what would happen if you change the operator separating the two conditions from and to or:

```
mysql> SELECT emp_id, fname, lname, start_date, title
    -> FROM employee
    -> WHERE title = 'Head Teller'
    ->   OR start_date > '2006-01-01';
+--------+----------+----------+------------+--------------------+
| emp_id | fname    | lname    | start_date | title              |
+--------+----------+----------+------------+--------------------+
|      2 | Susan    | Barker   | 2006-09-12 | Vice President     |
|      4 | Susan    | Hawthorne| 2006-04-24 | Operations Manager |
|      5 | John     | Gooding  | 2007-11-14 | Loan Manager       |
|      6 | Helen    | Fleming  | 2008-03-17 | Head Teller        |
|      7 | Chris    | Tucker   | 2008-09-15 | Teller             |
|      8 | Sarah    | Parker   | 2006-12-02 | Teller             |
|      9 | Jane     | Grossman | 2006-05-03 | Teller             |
|     10 | Paula    | Roberts  | 2006-07-27 | Head Teller        |
|     12 | Samantha | Jameson  | 2007-01-08 | Teller             |
|     13 | John     | Blake    | 2004-05-11 | Head Teller        |
```

```
|      14 | Cindy    | Mason    | 2006-08-09 | Teller      |
|      15 | Frank    | Portman  | 2007-04-01 | Teller      |
|      16 | Theresa  | Markham  | 2005-03-15 | Head Teller |
|      17 | Beth     | Fowler   | 2006-06-29 | Teller      |
|      18 | Rick     | Tulman   | 2006-12-12 | Teller      |
+--------+----------+----------+------------+--------------------+
15 rows in set (0.00 sec)
```

Looking at the output, you can see that all four head tellers are included in the result set, along with any other employee who started working for the bank after January 1, 2006. At least one of the two conditions is true for 15 of the 18 employees in the employee table. Thus, when you separate conditions using the and operator, *all* conditions must evaluate to true to be included in the result set; when you use or, however, only *one* of the conditions needs to evaluate to true for a row to be included.

So, what should you do if you need to use both and and or operators in your where clause? Glad you asked. You should use parentheses to group conditions together. The next query specifies that only those employees who are head tellers *and* began working for the company after January 1, 2006 *or* those employees who are tellers *and* began working after January 1, 2007 be included in the result set:

```
mysql> SELECT emp_id, fname, lname, start_date, title
    -> FROM employee
    -> WHERE (title = 'Head Teller' AND start_date > '2006-01-01')
    ->    OR (title = 'Teller' AND start_date > '2007-01-01');
+--------+----------+---------+------------+-------------+
| emp_id | fname    | lname   | start_date | title       |
+--------+----------+---------+------------+-------------+
|      6 | Helen    | Fleming | 2008-03-17 | Head Teller |
|      7 | Chris    | Tucker  | 2008-09-15 | Teller      |
|     10 | Paula    | Roberts | 2006-07-27 | Head Teller |
|     12 | Samantha | Jameson | 2007-01-08 | Teller      |
|     15 | Frank    | Portman | 2007-04-01 | Teller      |
+--------+----------+---------+------------+-------------+
5 rows in set (0.00 sec)
```

You should always use parentheses to separate groups of conditions when mixing different operators so that you, the database server, and anyone who comes along later to modify your code will be on the same page.

The group by and having Clauses

All the queries thus far have retrieved raw data without any manipulation. Sometimes, however, you will want to find trends in your data that will require the database server to cook the data a bit before you retrieve your result set. One such mechanism is the group by clause, which is used to group data by column values. For example, rather than looking at a list of employees and the departments to which they are assigned, you might want to look at a list of departments along with the number of employees assigned to each department. When using the group by clause, you may also use the having clause, which allows you to filter group data in the same way the `where` clause lets you filter raw data.

Here's a quick look at a query that counts all the employees in each department and returns the names of those departments having more than two employees:

```
mysql> SELECT d.name, count(e.emp_id) num_employees
    -> FROM department d INNER JOIN employee e
    ->   ON d.dept_id = e.dept_id
    -> GROUP BY d.name
    -> HAVING count(e.emp_id) > 2;
+----------------+---------------+
| name           | num_employees |
+----------------+---------------+
| Administration |             3 |
| Operations     |            14 |
+----------------+---------------+
2 rows in set (0.00 sec)
```

I wanted to briefly mention these two clauses so that they don't catch you by surprise later in the book, but they are a bit more advanced than the other four `select` clauses. Therefore, I ask that you wait until Chapter 8 for a full description of how and when to use `group by` and `having`.

The order by Clause

In general, the rows in a result set returned from a query are not in any particular order. If you want your result set in a particular order, you will need to instruct the server to sort the results using the `order by` clause:

The `order by` *clause is the mechanism for sorting your result set using either raw column data or expressions based on column data.*

For example, here's another look at an earlier query against the `account` table:

```
mysql> SELECT open_emp_id, product_cd
    -> FROM account;
+-------------+------------+
| open_emp_id | product_cd |
+-------------+------------+
|          10 | CHK        |
|          10 | SAV        |
|          10 | CD         |
|          10 | CHK        |
|          10 | SAV        |
|          13 | CHK        |
|          13 | MM         |
|           1 | CHK        |
|           1 | SAV        |
|           1 | MM         |
|          16 | CHK        |
|           1 | CHK        |
|           1 | CD         |
```

```
|          10 | CD       |
|          16 | CHK      |
|          16 | SAV      |
|           1 | CHK      |
|           1 | MM       |
|           1 | CD       |
|          16 | CHK      |
|          16 | BUS      |
|          10 | BUS      |
|          16 | CHK      |
|          13 | SBL      |
+-------------+----------+
24 rows in set (0.00 sec)
```

If you are trying to analyze data for each employee, it would be helpful to sort the results by the open_emp_id column; to do so, simply add this column to the **order by** clause:

```
mysql> SELECT open_emp_id, product_cd
    -> FROM account
    -> ORDER BY open_emp_id;
+-------------+------------+
| open_emp_id | product_cd |
+-------------+------------+
|           1 | CHK        |
|           1 | SAV        |
|           1 | MM         |
|           1 | CHK        |
|           1 | CD         |
|           1 | CHK        |
|           1 | MM         |
|           1 | CD         |
|          10 | CHK        |
|          10 | SAV        |
|          10 | CD         |
|          10 | CHK        |
|          10 | SAV        |
|          10 | CD         |
|          10 | BUS        |
|          13 | CHK        |
|          13 | MM         |
|          13 | SBL        |
|          16 | CHK        |
|          16 | CHK        |
|          16 | SAV        |
|          16 | CHK        |
|          16 | BUS        |
|          16 | CHK        |
+-------------+------------+
24 rows in set (0.00 sec)
```

It is now easier to see what types of accounts each employee opened. However, it might be even better if you could ensure that the account types were shown in the same order for each distinct employee; you can accomplish this by adding the product_cd column after the open_emp_id column in the **order by** clause:

```
mysql> SELECT open_emp_id, product_cd
    -> FROM account
    -> ORDER BY open_emp_id, product_cd;
+-------------+------------+
| open_emp_id | product_cd |
+-------------+------------+
|           1 | CD         |
|           1 | CD         |
|           1 | CHK        |
|           1 | CHK        |
|           1 | CHK        |
|           1 | MM         |
|           1 | MM         |
|           1 | SAV        |
|          10 | BUS        |
|          10 | CD         |
|          10 | CD         |
|          10 | CHK        |
|          10 | CHK        |
|          10 | SAV        |
|          10 | SAV        |
|          13 | CHK        |
|          13 | MM         |
|          13 | SBL        |
|          16 | BUS        |
|          16 | CHK        |
|          16 | CHK        |
|          16 | CHK        |
|          16 | CHK        |
|          16 | SAV        |
+-------------+------------+
24 rows in set (0.00 sec)
```

The result set has now been sorted first by employee ID and then by account type. The order in which columns appear in your order by clause does make a difference.

Ascending Versus Descending Sort Order

When sorting, you have the option of specifying *ascending* or *descending* order via the asc and desc keywords. The default is ascending, so you will need to add the desc keyword, only if you want to use a descending sort. For example, the following query lists all accounts sorted by available balance with the highest balance listed at the top:

```
mysql> SELECT account_id, product_cd, open_date, avail_balance
    -> FROM account
    -> ORDER BY avail_balance DESC;
+------------+------------+------------+---------------+
| account_id | product_cd | open_date  | avail_balance |
+------------+------------+------------+---------------+
|         29 | SBL        | 2004-02-22 |      50000.00 |
|         28 | CHK        | 2003-07-30 |      38552.05 |
|         24 | CHK        | 2002-09-30 |      23575.12 |
|         15 | CD         | 2004-12-28 |      10000.00 |
|         27 | BUS        | 2004-03-22 |       9345.55 |
```

```
|        22  |  MM         |  2004-10-28  |      9345.55  |
|        12  |  MM         |  2004-09-30  |      5487.09  |
|        17  |  CD         |  2004-01-12  |      5000.00  |
|        18  |  CHK        |  2001-05-23  |      3487.19  |
|         3  |  CD         |  2004-06-30  |      3000.00  |
|         4  |  CHK        |  2001-03-12  |      2258.02  |
|        13  |  CHK        |  2004-01-27  |      2237.97  |
|         8  |  MM         |  2002-12-15  |      2212.50  |
|        23  |  CD         |  2004-06-30  |      1500.00  |
|         1  |  CHK        |  2000-01-15  |      1057.75  |
|         7  |  CHK        |  2002-11-23  |      1057.75  |
|        11  |  SAV        |  2000-01-15  |       767.77  |
|        10  |  CHK        |  2003-09-12  |       534.12  |
|         2  |  SAV        |  2000-01-15  |       500.00  |
|        19  |  SAV        |  2001-05-23  |       387.99  |
|         5  |  SAV        |  2001-03-12  |       200.00  |
|        21  |  CHK        |  2003-07-30  |       125.67  |
|        14  |  CHK        |  2002-08-24  |       122.37  |
|        25  |  BUS        |  2002-10-01  |         0.00  |
+------------+------------+------------+---------------+
24 rows in set (0.05 sec)
```

Descending sorts are commonly used for ranking queries, such as "show me the top five account balances." MySQL includes a `limit` clause that allows you to sort your data and then discard all but the first *X* rows; see Appendix B for a discussion of the `limit` clause, along with other non-ANSI extensions.

Sorting via Expressions

Sorting your results using column data is all well and good, but sometimes you might need to sort by something that is not stored in the database, and possibly doesn't appear anywhere in your query. You can add an expression to your `order by` clause to handle such situations. For example, perhaps you would like to sort your customer data by the last three digits of the customer's federal ID number (which is either a Social Security number for individuals or a corporate ID for businesses):

```
mysql> SELECT cust_id, cust_type_cd, city, state, fed_id
    -> FROM customer
    -> ORDER BY RIGHT(fed_id, 3);
+---------+--------------+------------+-------+-------------+
| cust_id | cust_type_cd | city       | state | fed_id      |
+---------+--------------+------------+-------+-------------+
|       1 | I            | Lynnfield  | MA    | 111-11-1111 |
|      10 | B            | Salem      | NH    | 04-1111111  |
|       2 | I            | Woburn     | MA    | 222-22-2222 |
|      11 | B            | Wilmington | MA    | 04-2222222  |
|       3 | I            | Quincy     | MA    | 333-33-3333 |
|      12 | B            | Salem      | NH    | 04-3333333  |
|      13 | B            | Quincy     | MA    | 04-4444444  |
|       4 | I            | Waltham    | MA    | 444-44-4444 |
|       5 | I            | Salem      | NH    | 555-55-5555 |
|       6 | I            | Waltham    | MA    | 666-66-6666 |
|       7 | I            | Wilmington | MA    | 777-77-7777 |
```

```
|       8 | I              | Salem       | NH    | 888-88-8888 |
|       9 | I              | Newton      | MA    | 999-99-9999 |
+---------+----------------+-------------+-------+-------------+
13 rows in set (0.24 sec)
```

This query uses the built-in function `right()` to extract the last three characters of the fed_id column and then sorts the rows based on this value.

Sorting via Numeric Placeholders

If you are sorting using the columns in your `select` clause, you can opt to reference the columns by their *position* in the `select` clause rather than by name. For example, if you want to sort using the second and fifth columns returned by a query, you could do the following:

```
mysql> SELECT emp_id, title, start_date, fname, lname
    -> FROM employee
    -> ORDER BY 2, 5;
+---------+--------------------+------------+----------+-----------+
| emp_id  | title              | start_date | fname    | lname     |
+---------+--------------------+------------+----------+-----------+
|      13 | Head Teller        | 2004-05-11 | John     | Blake     |
|       6 | Head Teller        | 2008-03-17 | Helen    | Fleming   |
|      16 | Head Teller        | 2005-03-15 | Theresa  | Markham   |
|      10 | Head Teller        | 2006-07-27 | Paula    | Roberts   |
|       5 | Loan Manager       | 2007-11-14 | John     | Gooding   |
|       4 | Operations Manager | 2006-04-24 | Susan    | Hawthorne |
|       1 | President          | 2005-06-22 | Michael  | Smith     |
|      17 | Teller             | 2006-06-29 | Beth     | Fowler    |
|       9 | Teller             | 2006-05-03 | Jane     | Grossman  |
|      12 | Teller             | 2007-01-08 | Samantha | Jameson   |
|      14 | Teller             | 2006-08-09 | Cindy    | Mason     |
|       8 | Teller             | 2006-12-02 | Sarah    | Parker    |
|      15 | Teller             | 2007-04-01 | Frank    | Portman   |
|       7 | Teller             | 2008-09-15 | Chris    | Tucker    |
|      18 | Teller             | 2006-12-12 | Rick     | Tulman    |
|      11 | Teller             | 2004-10-23 | Thomas   | Ziegler   |
|       3 | Treasurer          | 2005-02-09 | Robert   | Tyler     |
|       2 | Vice President     | 2006-09-12 | Susan    | Barker    |
+---------+--------------------+------------+----------+-----------+
18 rows in set (0.00 sec)
```

You might want to use this feature sparingly, since adding a column to the `select` clause without changing the numbers in the `order by` clause can lead to unexpected results. Personally, I may reference columns positionally when writing ad hoc queries, but I always reference columns by name when writing code.

Test Your Knowledge

The following exercises are designed to strengthen your understanding of the select statement and its various clauses. Please see Appendix C for solutions.

Exercise 3-1

Retrieve the employee ID, first name, and last name for all bank employees. Sort by last name and then by first name.

SELECT emp_id, fname, lname

FROM employee

ORDER BY lname, fname ;

Exercise 3-2

Retrieve the account ID, customer ID, and available balance for all accounts whose status equals 'ACTIVE' and whose available balance is greater than $2,500.

SELECT account_id, cust_id, avail_balance

FROM account

WHERE status = 'ACTIVE'

AND avail_balance > 2500

Exercise 3-3

Write a query against the account table that returns the IDs of the employees who opened the accounts (use the account.open_emp_id column). Include a single row for each distinct employee.

SELECT DISTINCT open_emp_id

FROM account

Exercise 3-4

Fill in the blanks (denoted by <#>) for this multi-data-set query to achieve the results shown here:

```
mysql> SELECT p.product_cd, a.cust_id, a.avail_balance
    -> FROM product p INNER JOIN account <1>
    ->    ON p.product_cd = <2>
    -> WHERE p.<3> = 'ACCOUNT'
    -> ORDER BY <4>, <5>;
+------------+---------+---------------+
| product_cd | cust_id | avail_balance |
+------------+---------+---------------+
| CD         |       1 |       3000.00 |
| CD         |       6 |      10000.00 |
| CD         |       7 |       5000.00 |
| CD         |       9 |       1500.00 |
| CHK        |       1 |       1057.75 |
| CHK        |       2 |       2258.02 |
| CHK        |       3 |       1057.75 |
| CHK        |       4 |        534.12 |
| CHK        |       5 |       2237.97 |
| CHK        |       6 |        122.37 |
| CHK        |       8 |       3487.19 |
| CHK        |       9 |        125.67 |
| CHK        |      10 |      23575.12 |
| CHK        |      12 |      38552.05 |
| MM         |       3 |       2212.50 |
```

1 = a

2 = a.product_cd

3 = p.product_type_cd

4 = p.product_cd

5 = a.avail_balance

```
| MM             |       4 |      5487.09 |
| MM             |       9 |      9345.55 |
| SAV            |       1 |       500.00 |
| SAV            |       2 |       200.00 |
| SAV            |       4 |       767.77 |
| SAV            |       8 |       387.99 |
+----------------+---------+--------------+
21 rows in set (0.09 sec)
```

CHAPTER 4

Filtering

Sometimes you will want to work with every row in a table, such as:

- Purging all data from a table used to stage new data warehouse feeds
- Modifying all rows in a table after a new column has been added
- Retrieving all rows from a message queue table

In cases like these, your SQL statements won't need to have a `where` clause, since you don't need to exclude any rows from consideration. Most of the time, however, you will want to narrow your focus to a subset of a table's rows. Therefore, all the SQL data statements (except the `insert` statement) include an optional `where` clause to house *filter conditions* used to restrict the number of rows acted on by the SQL statement. Additionally, the `select` statement includes a `having` clause in which filter conditions pertaining to grouped data may be included. This chapter explores the various types of filter conditions that you can employ in the `where` clauses of `select`, `update`, and `delete` statements; we explore the use of filter conditions in the `having` clause of a `select` statement in Chapter 8.

Condition Evaluation

A `where` clause may contain one or more *conditions*, separated by the operators `and` and `or`. If multiple conditions are separated only by the `and` operator, then all the conditions must evaluate to `true` for the row to be included in the result set. Consider the following `where` clause:

```
WHERE title = 'Teller' AND start_date < '2007-01-01'
```

Given these two conditions, only tellers who began working for the bank prior to 2007 will be included (or, to look at it another way, any employee who is either not a teller or began working for the bank in 2007 or later will be removed from consideration). Although this example uses only two conditions, no matter how many conditions are in your `where` clause, if they are separated by the `and` operator they must *all* evaluate to `true` for the row to be included in the result set.

If all conditions in the where clause are separated by the or operator, however, only *one* of the conditions must evaluate to true for the row to be included in the result set. Consider the following two conditions:

```
WHERE title = 'Teller' OR start_date < '2007-01-01'
```

There are now various ways for a given employee row to be included in the result set:

- The employee is a teller and was employed prior to 2007.
- The employee is a teller and was employed after January 1, 2007.
- The employee is something other than a teller but was employed prior to 2007.

Table 4-1 shows the possible outcomes for a where clause containing two conditions separated by the or operator.

Table 4-1. Two-condition evaluation using or

Intermediate result	Final result
WHERE true OR true	True
WHERE true OR false	True
WHERE false OR true	True
WHERE false OR false	False

In the case of the preceding example, the only way for a row to be excluded from the result set is if the employee is not a teller and was employed on or after January 1, 2007.

Using Parentheses

If your where clause includes three or more conditions using both the and and or operators, you should use parentheses to make your intent clear, both to the database server and to anyone else reading your code. Here's a where clause that extends the previous example by checking to make sure that the employee is still employed by the bank:

```
WHERE end_date IS NULL
  AND (title = 'Teller' OR start_date < '2007-01-01')
```

There are now three conditions; for a row to make it to the final result set, the first condition must evaluate to true, and either the second *or* third condition (or both) must evaluate to true. Table 4-2 shows the possible outcomes for this where clause.

Table 4-2. Three-condition evaluation using and, or

Intermediate result	Final result
WHERE true AND (true OR true)	True
WHERE true AND (true OR false)	True
WHERE true AND (false OR true)	True
WHERE true AND (false OR false)	False

Intermediate result	Final result
`WHERE false AND (true OR true)`	False
`WHERE false AND (true OR false)`	False
`WHERE false AND (false OR true)`	False
`WHERE false AND (false OR false)`	False

As you can see, the more conditions you have in your **where** clause, the more combinations there are for the server to evaluate. In this case, only three of the eight combinations yield a final result of **true**.

Using the not Operator

Hopefully, the previous three-condition example is fairly easy to understand. Consider the following modification, however:

```
WHERE end_date IS NULL
  AND NOT (title = 'Teller' OR start_date < '2007-01-01')
```

Did you spot the change from the previous example? I added the **not** operator after the **and** operator on the second line. Now, instead of looking for nonterminated employees who either are tellers or began working for the bank prior to 2007, I am looking for nonterminated employees who both are nontellers and began working for the bank in 2007 or later. Table 4-3 shows the possible outcomes for this example.

Table 4-3. Three-condition evaluation using and, or, and not

Intermediate result	Final result
`WHERE true AND NOT (true OR true)`	False
`WHERE true AND NOT (true OR false)`	False
`WHERE true AND NOT (false OR true)`	False
`WHERE true AND NOT (false OR false)`	True
`WHERE false AND NOT (true OR true)`	False
`WHERE false AND NOT (true OR false)`	False
`WHERE false AND NOT (false OR true)`	False
`WHERE false AND NOT (false OR false)`	False

While it is easy for the database server to handle, it is typically difficult for a person to evaluate a **where** clause that includes the **not** operator, which is why you won't encounter it very often. In this case, you can rewrite the **where** clause to avoid using the **not** operator:

```
WHERE end_date IS NULL
  AND title != 'Teller' AND start_date >= '2007-01-01'
```

While I'm sure that the server doesn't have a preference, you probably have an easier time understanding this version of the where clause.

Building a Condition

Now that you have seen how the server evaluates multiple conditions, let's take a step back and look at what comprises a single condition. A condition is made up of one or more *expressions* coupled with one or more *operators*. An expression can be any of the following:

- A number
- A column in a table or view
- A string literal, such as 'Teller'
- A built-in function, such as concat('Learning', ' ', 'SQL')
- A subquery
- A list of expressions, such as ('Teller', 'Head Teller', 'Operations Manager')

The operators used within conditions include:

- Comparison operators, such as =, !=, <, >, <>, LIKE, IN, and BETWEEN
- Arithmetic operators, such as +, -, *, and /

The following section demonstrates how you can combine these expressions and operators to manufacture the various types of conditions.

Condition Types

There are many different ways to filter out unwanted data. You can look for specific values, sets of values, or ranges of values to include or exclude, or you can use various pattern-searching techniques to look for partial matches when dealing with string data. The next four subsections explore each of these condition types in detail.

Equality Conditions

A large percentage of the filter conditions that you write or come across will be of the form 'column = expression' as in:

```
title = 'Teller'
fed_id = '111-11-1111'
amount = 375.25
dept_id = (SELECT dept_id FROM department WHERE name = 'Loans')
```

Conditions such as these are called *equality conditions* because they equate one expression to another. The first three examples equate a column to a literal (two strings and a number), and the fourth example equates a column to the value returned from a subquery. The following query uses two equality conditions; one in the on clause (a join condition), and the other in the where clause (a filter condition):

```
mysql> SELECT pt.name product_type, p.name product
    -> FROM product p INNER JOIN product_type pt
    ->   ON p.product_type_cd = pt.product_type_cd
    -> WHERE pt.name = 'Customer Accounts';
+-------------------+------------------------+
| product_type      | product                |
+-------------------+------------------------+
| Customer Accounts | certificate of deposit |
| Customer Accounts | checking account       |
| Customer Accounts | money market account   |
| Customer Accounts | savings account        |
+-------------------+------------------------+
4 rows in set (0.08 sec)
```

This query shows all products that are *customer account* types.

Inequality conditions

Another fairly common type of condition is the *inequality condition*, which asserts that two expressions are *not* equal. Here's the previous query with the filter condition in the where clause changed to an inequality condition:

```
mysql> SELECT pt.name product_type, p.name product
    -> FROM product p INNER JOIN product_type pt
    ->   ON p.product_type_cd = pt.product_type_cd
    -> WHERE pt.name <> 'Customer Accounts';
+-----------------------------+------------------------+
| product_type                | product                |
+-----------------------------+------------------------+
| Individual and Business Loans | auto loan            |
| Individual and Business Loans | business line of credit |
| Individual and Business Loans | home mortgage        |
| Individual and Business Loans | small business loan  |
+-----------------------------+------------------------+
4 rows in set (0.00 sec)
```

This query shows all products that are *not* customer account types. When building inequality conditions, you may choose to use either the != or <> operator.

Data modification using equality conditions

Equality/inequality conditions are commonly used when modifying data. For example, let's say that the bank has a policy of removing old account rows once per year. Your task is to remove rows from the account table that were closed in 2002. Here's one way to tackle it:

```
DELETE FROM account
WHERE status = 'CLOSED' AND YEAR(close_date) = 2002;
```

This statement includes two equality conditions: one to find only closed accounts, and another to check for those accounts closed in 2002.

When crafting examples of `delete` and `update` statements, I try to write each statement such that no rows are modified. That way, when you execute the statements, your data will remain unchanged, and your output from `select` statements will always match that shown in this book.

Since MySQL sessions are in auto-commit mode by default (see Chapter 12), you would not be able to roll back (undo) any changes made to the example data if one of my statements modified the data. You may, of course, do whatever you want with the example data, including wiping it clean and rerunning the scripts I have provided, but I try to leave it intact.

Range Conditions

Along with checking that an expression is equal to (or not equal to) another expression, you can build conditions that check whether an expression falls within a certain range. This type of condition is common when working with numeric or temporal data. Consider the following query:

```
mysql> SELECT emp_id, fname, lname, start_date
    -> FROM employee
    -> WHERE start_date < '2007-01-01';
+--------+---------+-----------+------------+
| emp_id | fname   | lname     | start_date |
+--------+---------+-----------+------------+
|      1 | Michael | Smith     | 2005-06-22 |
|      2 | Susan   | Barker    | 2006-09-12 |
|      3 | Robert  | Tyler     | 2005-02-09 |
|      4 | Susan   | Hawthorne | 2006-04-24 |
|      8 | Sarah   | Parker    | 2006-12-02 |
|      9 | Jane    | Grossman  | 2006-05-03 |
|     10 | Paula   | Roberts   | 2006-07-27 |
|     11 | Thomas  | Ziegler   | 2004-10-23 |
|     13 | John    | Blake     | 2004-05-11 |
|     14 | Cindy   | Mason     | 2006-08-09 |
|     16 | Theresa | Markham   | 2005-03-15 |
|     17 | Beth    | Fowler    | 2006-06-29 |
|     18 | Rick    | Tulman    | 2006-12-12 |
+--------+---------+-----------+------------+
13 rows in set (0.15 sec)
```

This query finds all employees hired prior to 2007. Along with specifying an upper limit for the start date, you may also want to specify a lower range for the start date:

```
mysql> SELECT emp_id, fname, lname, start_date
    -> FROM employee
    -> WHERE start_date < '2007-01-01'
    ->   AND start_date >= '2005-01-01';
+--------+---------+-----------+------------+
| emp_id | fname   | lname     | start_date |
+--------+---------+-----------+------------+
```

```
|       1 | Michael  | Smith      | 2005-06-22 |
|       2 | Susan    | Barker     | 2006-09-12 |
|       3 | Robert   | Tyler      | 2005-02-09 |
|       4 | Susan    | Hawthorne  | 2006-04-24 |
|       8 | Sarah    | Parker     | 2006-12-02 |
|       9 | Jane     | Grossman   | 2006-05-03 |
|      10 | Paula    | Roberts    | 2006-07-27 |
|      14 | Cindy    | Mason      | 2006-08-09 |
|      16 | Theresa  | Markham    | 2005-03-15 |
|      17 | Beth     | Fowler     | 2006-06-29 |
|      18 | Rick     | Tulman     | 2006-12-12 |
+--------+---------+-----------+------------+
11 rows in set (0.00 sec)
```

This version of the query retrieves all employees hired in 2005 or 2006.

The between operator

When you have *both* an upper and lower limit for your range, you may choose to use a single condition that utilizes the between operator rather than using two separate conditions, as in:

```
mysql> SELECT emp_id, fname, lname, start_date
    -> FROM employee
    -> WHERE start_date BETWEEN '2005-01-01' AND '2007-01-01';
+--------+---------+-----------+------------+
| emp_id | fname   | lname     | start_date |
+--------+---------+-----------+------------+
|      1 | Michael  | Smith      | 2005-06-22 |
|      2 | Susan    | Barker     | 2006-09-12 |
|      3 | Robert   | Tyler      | 2005-02-09 |
|      4 | Susan    | Hawthorne  | 2006-04-24 |
|      8 | Sarah    | Parker     | 2006-12-02 |
|      9 | Jane     | Grossman   | 2006-05-03 |
|     10 | Paula    | Roberts    | 2006-07-27 |
|     14 | Cindy    | Mason      | 2006-08-09 |
|     16 | Theresa  | Markham    | 2005-03-15 |
|     17 | Beth     | Fowler     | 2006-06-29 |
|     18 | Rick     | Tulman     | 2006-12-12 |
+--------+---------+-----------+------------+
11 rows in set (0.03 sec)
```

When using the between operator, there are a couple of things to keep in mind. You should always specify the lower limit of the range first (after between) and the upper limit of the range second (after and). Here's what happens if you mistakenly specify the upper limit first:

```
mysql> SELECT emp_id, fname, lname, start_date
    -> FROM employee
    -> WHERE start_date BETWEEN '2007-01-01' AND '2005-01-01';
Empty set (0.00 sec)
```

As you can see, no data is returned. This is because the server is, in effect, generating two conditions from your single condition using the <= and >= operators, as in:

```
mysql> SELECT emp_id, fname, lname, start_date
    -> FROM employee
    -> WHERE start_date >= '2007-01-01'
    ->   AND start_date <= '2005-01-01';
Empty set (0.00 sec)
```

Since it is impossible to have a date that is *both* greater than January 1, 2007 and less than January 1, 2005, the query returns an empty set. This brings me to the second pitfall when using between, which is to remember that your upper and lower limits are *inclusive*, meaning that the values you provide are included in the range limits. In this case, I want to specify 2005-01-01 as the lower end of the range and 2006-12-31 as the upper end, rather than 2007-01-01. Even though there probably weren't any employees who started working for the bank on New Year's Day 2007, it is best to specify exactly what you want.

Along with dates, you can also build conditions to specify ranges of numbers. Numeric ranges are fairly easy to grasp, as demonstrated by the following:

```
mysql> SELECT account_id, product_cd, cust_id, avail_balance
    -> FROM account
    -> WHERE avail_balance BETWEEN 3000 AND 5000;
+------------+------------+---------+---------------+
| account_id | product_cd | cust_id | avail_balance |
+------------+------------+---------+---------------+
|          3 | CD         |       1 |       3000.00 |
|         17 | CD         |       7 |       5000.00 |
|         18 | CHK        |       8 |       3487.19 |
+------------+------------+---------+---------------+
3 rows in set (0.10 sec)
```

All accounts with between $3,000 and $5,000 of an available balance are returned. Again, make sure that you specify the lower amount first.

String ranges

While ranges of dates and numbers are easy to understand, you can also build conditions that search for ranges of strings, which are a bit harder to visualize. Say, for example, you are searching for customers having a Social Security number that falls within a certain range. The format for a Social Security number is "XXX-XX-XXXX," where X is a number from 0 to 9, and you want to find every customer whose Social Security number lies between "500-00-0000" and "999-99-9999." Here's what the statement would look like:

```
mysql> SELECT cust_id, fed_id
    -> FROM customer
    -> WHERE cust_type_cd = 'I'
    ->   AND fed_id BETWEEN '500-00-0000' AND '999-99-9999';
+---------+-------------+
| cust_id | fed_id      |
+---------+-------------+
|       5 | 555-55-5555 |
|       6 | 666-66-6666 |
```

```
|         7 | 777-77-7777 |
|         8 | 888-88-8888 |
|         9 | 999-99-9999 |
+---------+-------------+
5 rows in set (0.01 sec)
```

To work with string ranges, you need to know the order of the characters within your character set (the order in which the characters within a character set are sorted is called a *collation*).

Membership Conditions

In some cases, you will not be restricting an expression to a single value or range of values, but rather to a finite set of values. For example, you might want to locate all accounts whose product code is either 'CHK', 'SAV', 'CD', or 'MM':

```
mysql> SELECT account_id, product_cd, cust_id, avail_balance
    -> FROM account
    -> WHERE product_cd = 'CHK' OR product_cd = 'SAV'
    ->   OR product_cd = 'CD' OR product_cd = 'MM';
+------------+------------+---------+---------------+
| account_id | product_cd | cust_id | avail_balance |
+------------+------------+---------+---------------+
|          1 | CHK        |       1 |       1057.75 |
|          2 | SAV        |       1 |        500.00 |
|          3 | CD         |       1 |       3000.00 |
|          4 | CHK        |       2 |       2258.02 |
|          5 | SAV        |       2 |        200.00 |
|          7 | CHK        |       3 |       1057.75 |
|          8 | MM         |       3 |       2212.50 |
|         10 | CHK        |       4 |        534.12 |
|         11 | SAV        |       4 |        767.77 |
|         12 | MM         |       4 |       5487.09 |
|         13 | CHK        |       5 |       2237.97 |
|         14 | CHK        |       6 |        122.37 |
|         15 | CD         |       6 |      10000.00 |
|         17 | CD         |       7 |       5000.00 |
|         18 | CHK        |       8 |       3487.19 |
|         19 | SAV        |       8 |        387.99 |
|         21 | CHK        |       9 |        125.67 |
|         22 | MM         |       9 |       9345.55 |
|         23 | CD         |       9 |       1500.00 |
|         24 | CHK        |      10 |      23575.12 |
|         28 | CHK        |      12 |      38552.05 |
+------------+------------+---------+---------------+
21 rows in set (0.28 sec)
```

While this where clause (four conditions or'd together) wasn't too tedious to generate, imagine if the set of expressions contained 10 or 20 members. For these situations, you can use the in operator instead:

```
SELECT account_id, product_cd, cust_id, avail_balance
FROM account
WHERE product_cd IN ('CHK','SAV','CD','MM');
```

With the in operator, you can write a single condition no matter how many expressions are in the set.

Using subqueries

Along with writing your own set of expressions, such as ('CHK','SAV','CD','MM'), you can use a subquery to generate a set for you on the fly. For example, all four product types used in the previous query have a product_type_cd of 'ACCOUNT', so why not use a subquery against the product table to retrieve the four product codes instead of explicitly naming them:

```
mysql> SELECT account_id, product_cd, cust_id, avail_balance
    -> FROM account
    -> WHERE product_cd IN (SELECT product_cd FROM product
    ->   WHERE product_type_cd = 'ACCOUNT');
+------------+------------+---------+---------------+
| account_id | product_cd | cust_id | avail_balance |
+------------+------------+---------+---------------+
|          3 | CD         |       1 |       3000.00 |
|         15 | CD         |       6 |      10000.00 |
|         17 | CD         |       7 |       5000.00 |
|         23 | CD         |       9 |       1500.00 |
|          1 | CHK        |       1 |       1057.75 |
|          4 | CHK        |       2 |       2258.02 |
|          7 | CHK        |       3 |       1057.75 |
|         10 | CHK        |       4 |        534.12 |
|         13 | CHK        |       5 |       2237.97 |
|         14 | CHK        |       6 |        122.37 |
|         18 | CHK        |       8 |       3487.19 |
|         21 | CHK        |       9 |        125.67 |
|         24 | CHK        |      10 |      23575.12 |
|         28 | CHK        |      12 |      38552.05 |
|          8 | MM         |       3 |       2212.50 |
|         12 | MM         |       4 |       5487.09 |
|         22 | MM         |       9 |       9345.55 |
|          2 | SAV        |       1 |        500.00 |
|          5 | SAV        |       2 |        200.00 |
|         11 | SAV        |       4 |        767.77 |
|         19 | SAV        |       8 |        387.99 |
+------------+------------+---------+---------------+
21 rows in set (0.11 sec)
```

The subquery returns a set of four values, and the main query checks to see whether the value of the product_cd column can be found in the set that the subquery returned.

Using not in

Sometimes you want to see whether a particular expression exists within a set of expressions, and sometimes you want to see whether the expression does *not* exist. For these situations, you can use the not in operator:

```
mysql> SELECT account_id, product_cd, cust_id, avail_balance
    -> FROM account
    -> WHERE product_cd NOT IN ('CHK','SAV','CD','MM');
+------------+------------+---------+---------------+
| account_id | product_cd | cust_id | avail_balance |
+------------+------------+---------+---------------+
|         25 | BUS        |      10 |          0.00 |
|         27 | BUS        |      11 |       9345.55 |
|         29 | SBL        |      13 |      50000.00 |
+------------+------------+---------+---------------+
3 rows in set (0.09 sec)
```

This query finds all accounts that are *not* checking, savings, certificate of deposit, or money market accounts.

Matching Conditions

So far, you have been introduced to conditions that identify an exact string, a range of strings, or a set of strings; the final condition type deals with partial string matches. You may, for example, want to find all employees whose last name begins with *T*. You could use a built-in function to strip off the first letter of the lname column, as in:

```
mysql> SELECT emp_id, fname, lname
    -> FROM employee
    -> WHERE LEFT(lname, 1) = 'T';
+--------+--------+--------+
| emp_id | fname  | lname  |
+--------+--------+--------+
|      3 | Robert | Tyler  |
|      7 | Chris  | Tucker |
|     18 | Rick   | Tulman |
+--------+--------+--------+
3 rows in set (0.01 sec)
```

While the built-in function left() does the job, it doesn't give you much flexibility. Instead, you can use wildcard characters to build search expressions, as demonstrated in the next section.

Using wildcards

When searching for partial string matches, you might be interested in:

- Strings beginning/ending with a certain character
- Strings beginning/ending with a substring
- Strings containing a certain character anywhere within the string
- Strings containing a substring anywhere within the string
- Strings with a specific format, regardless of individual characters

You can build search expressions to identify these and many other partial string matches by using the wildcard characters shown in Table 4-4.

Table 4-4. Wildcard characters

Wildcard character	Matches
_	Exactly one character
%	Any number of characters (including 0)

The underscore character takes the place of a single character, while the percent sign can take the place of a variable number of characters. When building conditions that utilize search expressions, you use the `like` operator, as in:

```
mysql> SELECT lname
    -> FROM employee
    -> WHERE lname LIKE '_a%e%';
+-----------+
| lname     |
+-----------+
| Barker    |
| Hawthorne |
| Parker    |
| Jameson   |
+-----------+
4 rows in set (0.00 sec)
```

The search expression in the previous example specifies strings containing an *a* in the second position and followed by an *e* at any other position in the string (including the last position). Table 4-5 shows some more search expressions and their interpretations.

Table 4-5. Sample search expressions

Search expression	Interpretation
F%	Strings beginning with *F*
%t	Strings ending with *t*
%bas%	Strings containing the substring 'bas'
_ _t_	Four-character strings with a *t* in the third position
_ _ _-_ _-_ _ _ _	11-character strings with dashes in the fourth and seventh positions

You could use the last example in Table 4-5 to find customers whose federal ID matches the format used for Social Security numbers, as in:

```
mysql> SELECT cust_id, fed_id
    -> FROM customer
    -> WHERE fed_id LIKE '___-__-____';
+---------+-------------+
| cust_id | fed_id      |
+---------+-------------+
|       1 | 111-11-1111 |
|       2 | 222-22-2222 |
|       3 | 333-33-3333 |
|       4 | 444-44-4444 |
|       5 | 555-55-5555 |
```

```
|        6 | 666-66-6666 |
|        7 | 777-77-7777 |
|        8 | 888-88-8888 |
|        9 | 999-99-9999 |
+----------+-------------+
9 rows in set (0.02 sec)
```

The wildcard characters work fine for building simple search expressions; if your needs are a bit more sophisticated, however, you can use multiple search expressions, as demonstrated by the following:

```
mysql> SELECT emp_id, fname, lname
    -> FROM employee
    -> WHERE lname LIKE 'F%' OR lname LIKE 'G%';
+--------+-------+----------+
| emp_id | fname | lname    |
+--------+-------+----------+
|      5 | John  | Gooding  |
|      6 | Helen | Fleming  |
|      9 | Jane  | Grossman |
|     17 | Beth  | Fowler   |
+--------+-------+----------+
4 rows in set (0.00 sec)
```

This query finds all employees whose last name begins with *F* or *G*.

Using regular expressions

If you find that the wildcard characters don't provide enough flexibility, you can use regular expressions to build search expressions. A regular expression is, in essence, a search expression on steroids. If you are new to SQL but have coded using programming languages such as Perl, then you might already be intimately familiar with regular expressions. If you have never used regular expressions, then you may want to consult Jeffrey E.F. Friedl's *Mastering Regular Expressions (http://oreilly.com/catalog/9780596528126/)* (O'Reilly), since it is far too large a topic to try to cover in this book.

Here's what the previous query (find all employees whose last name starts with *F* or *G*) would look like using the MySQL implementation of regular expressions:

```
mysql> SELECT emp_id, fname, lname
    -> FROM employee
    -> WHERE lname REGEXP '^[FG]';
+--------+-------+----------+
| emp_id | fname | lname    |
+--------+-------+----------+
|      5 | John  | Gooding  |
|      6 | Helen | Fleming  |
|      9 | Jane  | Grossman |
|     17 | Beth  | Fowler   |
+--------+-------+----------+
4 rows in set (0.00 sec)
```

The regexp operator takes a regular expression ('^[FG]' in this example) and applies it to the expression on the lefthand side of the condition (the column lname). The query now contains a single condition using a regular expression rather than two conditions using wildcard characters.

Oracle Database and Microsoft SQL Server also support regular expressions. With Oracle Database, you would use the `regexp_like` function instead of the `regexp` operator shown in the previous example, whereas SQL Server allows regular expressions to be used with the `like` operator.

Null: That Four-Letter Word

I put it off as long as I could, but it's time to broach a topic that tends to be met with fear, uncertainty, and dread: the `null` value. Null is the absence of a value; before an employee is terminated, for example, her `end_date` column in the `employee` table should be `null`. There is no value that can be assigned to the `end_date` column that would make sense in this situation. Null is a bit slippery, however, as there are various flavors of `null`:

Not applicable
> Such as the employee ID column for a transaction that took place at an ATM machine

Value not yet known
> Such as when the federal ID is not known at the time a customer row is created

Value undefined
> Such as when an account is created for a product that has not yet been added to the database

 Some theorists argue that there should be a different expression to cover each of these (and more) situations, but most practitioners would agree that having multiple `null` values would be far too confusing.

When working with `null`, you should remember:

* An expression can *be* null, but it can never *equal* null.
* Two nulls are never equal to each other.

To test whether an expression is `null`, you need to use the `is null` operator, as demonstrated by the following:

```
mysql> SELECT emp_id, fname, lname, superior_emp_id
    -> FROM employee
    -> WHERE superior_emp_id IS NULL;
+--------+--------+-------+-----------------+
| emp_id | fname  | lname | superior_emp_id |
+--------+--------+-------+-----------------+
|      1 | Michael| Smith |            NULL |
+--------+--------+-------+-----------------+
1 row in set (0.00 sec)
```

This query returns all employees who do not have a boss (wouldn't that be nice?). Here's the same query using = null instead of is null:

```
mysql> SELECT emp_id, fname, lname, superior_emp_id
    -> FROM employee
    -> WHERE superior_emp_id = NULL;
Empty set (0.01 sec)
```

As you can see, the query parses and executes but does not return any rows. This is a common mistake made by inexperienced SQL programmers, and the database server will not alert you to your error, so be careful when constructing conditions that test for null.

If you want to see whether a value has been assigned to a column, you can use the is not null operator, as in:

```
mysql> SELECT emp_id, fname, lname, superior_emp_id
    -> FROM employee
    -> WHERE superior_emp_id IS NOT NULL;
+--------+----------+-----------+-----------------+
| emp_id | fname    | lname     | superior_emp_id |
+--------+----------+-----------+-----------------+
|      2 | Susan    | Barker    |               1 |
|      3 | Robert   | Tyler     |               1 |
|      4 | Susan    | Hawthorne |               3 |
|      5 | John     | Gooding   |               4 |
|      6 | Helen    | Fleming   |               4 |
|      7 | Chris    | Tucker    |               6 |
|      8 | Sarah    | Parker    |               6 |
|      9 | Jane     | Grossman  |               6 |
|     10 | Paula    | Roberts   |               4 |
|     11 | Thomas   | Ziegler   |              10 |
|     12 | Samantha | Jameson   |              10 |
|     13 | John     | Blake     |               4 |
|     14 | Cindy    | Mason     |              13 |
|     15 | Frank    | Portman   |              13 |
|     16 | Theresa  | Markham   |               4 |
|     17 | Beth     | Fowler    |              16 |
|     18 | Rick     | Tulman    |              16 |
+--------+----------+-----------+-----------------+
17 rows in set (0.00 sec)
```

This version of the query returns the other 17 employees who, unlike Michael Smith, have a boss.

Before putting null aside for a while, it would be helpful to investigate one more potential pitfall. Suppose that you have been asked to identify all employees who are *not* managed by Helen Fleming (whose employee ID is 6). Your first instinct might be to do the following:

```
mysql> SELECT emp_id, fname, lname, superior_emp_id
    -> FROM employee
    -> WHERE superior_emp_id != 6;
```

```
+--------+----------+-----------+-----------------+
| emp_id | fname    | lname     | superior_emp_id |
+--------+----------+-----------+-----------------+
|      2 | Susan    | Barker    |               1 |
|      3 | Robert   | Tyler     |               1 |
|      4 | Susan    | Hawthorne |               3 |
|      5 | John     | Gooding   |               4 |
|      6 | Helen    | Fleming   |               4 |
|     10 | Paula    | Roberts   |               4 |
|     11 | Thomas   | Ziegler   |              10 |
|     12 | Samantha | Jameson   |              10 |
|     13 | John     | Blake     |               4 |
|     14 | Cindy    | Mason     |              13 |
|     15 | Frank    | Portman   |              13 |
|     16 | Theresa  | Markham   |               4 |
|     17 | Beth     | Fowler    |              16 |
|     18 | Rick     | Tulman    |              16 |
+--------+----------+-----------+-----------------+
14 rows in set (0.00 sec)
```

While it is true that these 14 employees do not work for Helen Fleming, if you look carefully at the data, you will see that there is one more employee who doesn't work for Helen who is not listed here. That employee is Michael Smith, and his superior_emp_id column is null (because he's the big cheese). To answer the question correctly, therefore, you need to account for the possibility that some rows might contain a null in the superior_emp_id column:

```
mysql> SELECT emp_id, fname, lname, superior_emp_id
    -> FROM employee
    -> WHERE superior_emp_id != 6 OR superior_emp_id IS NULL;
+--------+----------+-----------+-----------------+
| emp_id | fname    | lname     | superior_emp_id |
+--------+----------+-----------+-----------------+
|      1 | Michael  | Smith     |            NULL |
|      2 | Susan    | Barker    |               1 |
|      3 | Robert   | Tyler     |               1 |
|      4 | Susan    | Hawthorne |               3 |
|      5 | John     | Gooding   |               4 |
|      6 | Helen    | Fleming   |               4 |
|     10 | Paula    | Roberts   |               4 |
|     11 | Thomas   | Ziegler   |              10 |
|     12 | Samantha | Jameson   |              10 |
|     13 | John     | Blake     |               4 |
|     14 | Cindy    | Mason     |              13 |
|     15 | Frank    | Portman   |              13 |
|     16 | Theresa  | Markham   |               4 |
|     17 | Beth     | Fowler    |              16 |
|     18 | Rick     | Tulman    |              16 |
+--------+----------+-----------+-----------------+
15 rows in set (0.00 sec)
```

The result set now includes all 15 employees who don't work for Helen. When working with a database that you are not familiar with, it is a good idea to find out which columns in a table allow `nulls` so that you can take appropriate measures with your filter conditions to keep data from slipping through the cracks.

Test Your Knowledge

The following exercises test your understanding of filter conditions. Please see Appendix C for solutions.

The following transaction data is used for the first two exercises:

Txn_id	Txn_date	Account_id	Txn_type_cd	Amount
1	2005-02-22	101	CDT	1000.00
2	2005-02-23	102	CDT	525.75
3	2005-02-24	101	DBT	100.00
4	2005-02-24	103	CDT	55
5	2005-02-25	101	DBT	50
6	2005-02-25	103	DBT	25
7	2005-02-25	102	CDT	125.37
8	2005-02-26	103	DBT	10
9	2005-02-27	101	CDT	75

Exercise 4-1

Which of the transaction IDs would be returned by the following filter conditions?

```
txn_date < '2005-02-26' AND (txn_type_cd = 'DBT' OR amount > 100)
```

1 2 3 5 6 7

Exercise 4-2

Which of the transaction IDs would be returned by the following filter conditions?

```
account_id IN (101,103) AND NOT (txn_type_cd = 'DBT' OR amount > 100)
```

4 9

Exercise 4-3

Construct a query that retrieves all accounts opened in 2002.

SELECT account_id, open_date
FROM account
WHERE open_date >= '2002-01-01' AND < 2003-01-01

OTHER WHERE OPTIONS:
YEAR (open_date) = 2002
open_date BETWEEN 2002-01-01
AND 2002-12-31

Exercise 4-4

Construct a query that finds all nonbusiness customers whose last name contains an *a* in the second position and an *e* anywhere after the *a*.

SELECT cust_id, fname, lname
FROM individual

AND lname LIKE '_a%e%';

CHAPTER 5

Querying Multiple Tables

Back in Chapter 2, I demonstrated how related concepts are broken into separate pieces through a process known as normalization. The end result of this exercise was two tables: `person` and `favorite_food`. If, however, you want to generate a single report showing a person's name, address, *and* favorite foods, you will need a mechanism to bring the data from these two tables back together again; this mechanism is known as a *join*, and this chapter concentrates on the simplest and most common join, the *inner join*. Chapter 10 demonstrates all of the different join types.

What Is a Join?

Queries against a single table are certainly not rare, but you will find that most of your queries will require two, three, or even more tables. To illustrate, let's look at the definitions for the `employee` and `department` tables and then define a query that retrieves data from both tables:

```
mysql> DESC employee;
+-------------------+----------------------+------+-----+---------+
| Field             | Type                 | Null | Key | Default |
+-------------------+----------------------+------+-----+---------+
| emp_id            | smallint(5) unsigned | NO   | PRI | NULL    |
| fname             | varchar(20)          | NO   |     | NULL    |
| lname             | varchar(20)          | NO   |     | NULL    |
| start_date        | date                 | NO   |     | NULL    |
| end_date          | date                 | YES  |     | NULL    |
| superior_emp_id   | smallint(5) unsigned | YES  | MUL | NULL    |
| dept_id           | smallint(5) unsigned | YES  | MUL | NULL    |
| title             | varchar(20)          | YES  |     | NULL    |
| assigned_branch_id | smallint(5) unsigned | YES | MUL | NULL    |
+-------------------+----------------------+------+-----+---------+
9 rows in set (0.11 sec)

 mysql> DESC department;
+---------+----------------------+------+-----+---------+
| Field   | Type                 | Null | Key | Default |
+---------+----------------------+------+-----+---------+
| dept_id | smallint(5) unsigned | No   | PRI | NULL    |
```

```
| name      | varchar(20)          | No  |     | NULL    |
+---------+----------------------+------+-----+---------+
2 rows in set (0.03 sec)
```

Let's say you want to retrieve the first and last names of each employee along with the name of the department to which each employee is assigned. Your query will therefore need to retrieve the employee.fname, employee.lname, and department.name columns. But how can you retrieve data from both tables in the same query? The answer lies in the employee.dept_id column, which holds the ID of the department to which each employee is assigned (in more formal terms, the employee.dept_id column is the *foreign key* to the department table). The query, which you will see shortly, instructs the server to use the employee.dept_id column as the *bridge* between the employee and department tables, thereby allowing columns from both tables to be included in the query's result set. This type of operation is known as a join.

Cartesian Product

The easiest way to start is to put the employee and department tables into the from clause of a query and see what happens. Here's a query that retrieves the employee's first and last names along with the department name, with a from clause naming both tables separated by the join keyword:

```
mysql> SELECT e.fname, e.lname, d.name
    -> FROM employee e JOIN department d;
+----------+-----------+----------------+
| fname    | lname     | name           |
+----------+-----------+----------------+
| Michael  | Smith     | Operations     |
| Michael  | Smith     | Loans          |
| Michael  | Smith     | Administration |
| Susan    | Barker    | Operations     |
| Susan    | Barker    | Loans          |
| Susan    | Barker    | Administration |
| Robert   | Tyler     | Operations     |
| Robert   | Tyler     | Loans          |
| Robert   | Tyler     | Administration |
| Susan    | Hawthorne | Operations     |
| Susan    | Hawthorne | Loans          |
| Susan    | Hawthorne | Administration |
| John     | Gooding   | Operations     |
| John     | Gooding   | Loans          |
| John     | Gooding   | Administration |
| Helen    | Fleming   | Operations     |
| Helen    | Fleming   | Loans          |
| Helen    | Fleming   | Administration |
| Chris    | Tucker    | Operations     |
| Chris    | Tucker    | Loans          |
| Chris    | Tucker    | Administration |
| Sarah    | Parker    | Operations     |
| Sarah    | Parker    | Loans          |
| Sarah    | Parker    | Administration |
| Jane     | Grossman  | Operations     |
```

```
| Jane     | Grossman | Loans          |
| Jane     | Grossman | Administration |
| Paula    | Roberts  | Operations     |
| Paula    | Roberts  | Loans          |
| Paula    | Roberts  | Administration |
| Thomas   | Ziegler  | Operations     |
| Thomas   | Ziegler  | Loans          |
| Thomas   | Ziegler  | Administration |
| Samantha | Jameson  | Operations     |
| Samantha | Jameson  | Loans          |
| Samantha | Jameson  | Administration |
| John     | Blake    | Operations     |
| John     | Blake    | Loans          |
| John     | Blake    | Administration |
| Cindy    | Mason    | Operations     |
| Cindy    | Mason    | Loans          |
| Cindy    | Mason    | Administration |
| Frank    | Portman  | Operations     |
| Frank    | Portman  | Loans          |
| Frank    | Portman  | Administration |
| Theresa  | Markham  | Operations     |
| Theresa  | Markham  | Loans          |
| Theresa  | Markham  | Administration |
| Beth     | Fowler   | Operations     |
| Beth     | Fowler   | Loans          |
| Beth     | Fowler   | Administration |
| Rick     | Tulman   | Operations     |
| Rick     | Tulman   | Loans          |
| Rick     | Tulman   | Administration |
+----------+----------+----------------+
54 rows in set (0.23 sec)
```

Hmmm...there are only 18 employees and 3 different departments, so how did the result set end up with 54 rows? Looking more closely, you can see that the set of 18 employees is repeated three times, with all the data identical except for the department name. Because the query didn't specify *how* the two tables should be joined, the database server generated the *Cartesian product*, which is *every* permutation of the two tables (18 employees × 3 departments = 54 permutations). This type of join is known as a *cross join*, and it is rarely used (on purpose, at least). Cross joins are one of the join types that we study in Chapter 10.

Inner Joins

To modify the previous query so that only 18 rows are included in the result set (one for each employee), you need to describe how the two tables are related. Earlier, I showed that the employee.dept_id column serves as the link between the two tables, so this information needs to be added to the on subclause of the from clause:

```
mysql> SELECT e.fname, e.lname, d.name
    -> FROM employee e JOIN department d
    ->   ON e.dept_id = d.dept_id;
```

```
+-----------+-----------+----------------+
| fname     | lname     | name           |
+-----------+-----------+----------------+
| Michael   | Smith     | Administration |
| Susan     | Barker    | Administration |
| Robert    | Tyler     | Administration |
| Susan     | Hawthorne | Operations     |
| John      | Gooding   | Loans          |
| Helen     | Fleming   | Operations     |
| Chris     | Tucker    | Operations     |
| Sarah     | Parker    | Operations     |
| Jane      | Grossman  | Operations     |
| Paula     | Roberts   | Operations     |
| Thomas    | Ziegler   | Operations     |
| Samantha  | Jameson   | Operations     |
| John      | Blake     | Operations     |
| Cindy     | Mason     | Operations     |
| Frank     | Portman   | Operations     |
| Theresa   | Markham   | Operations     |
| Beth      | Fowler    | Operations     |
| Rick      | Tulman    | Operations     |
+-----------+-----------+----------------+
18 rows in set (0.00 sec)
```

Instead of 54 rows, you now have the expected 18 rows due to the addition of the on subclause, which instructs the server to join the employee and department tables by using the dept_id column to traverse from one table to the other. For example, Susan Hawthorne's row in the employee table contains a value of 1 in the dept_id column (not shown in the example). The server uses this value to look up the row in the department table having a value of 1 in its dept_id column and then retrieves the value 'Operations' from the name column in that row.

If a value exists for the dept_id column in one table but *not* the other, then the join fails for the rows containing that value and those rows are excluded from the result set. This type of join is known as an *inner join*, and it is the most commonly used type of join. To clarify, if the department table contains a fourth row for the marketing department, but no employees have been assigned to that department, then the marketing department would not be included in the result set. Likewise, if some of the employees had been assigned to department ID 99, which doesn't exist in the department table, then these employees would be left out of the result set. If you want to include all rows from one table or the other regardless of whether a match exists, you need to specify an *outer join*, but we cover this later in the book.

In the previous example, I did not specify in the from clause which type of join to use. However, when you wish to join two tables using an inner join, you should explicitly specify this in your from clause; here's the same example, with the addition of the join type (note the keyword INNER):

```
mysql> SELECT e.fname, e.lname, d.name
    -> FROM employee e INNER JOIN department d
```

```
    ->    ON e.dept_id = d.dept_id;
+-----------+-----------+----------------+
| fname     | lname     | name           |
+-----------+-----------+----------------+
| Michael   | Smith     | Administration |
| Susan     | Barker    | Administration |
| Robert    | Tyler     | Administration |
| Susan     | Hawthorne | Operations     |
| John      | Gooding   | Loans          |
| Helen     | Fleming   | Operations     |
| Chris     | Tucker    | Operations     |
| Sarah     | Parker    | Operations     |
| Jane      | Grossman  | Operations     |
| Paula     | Roberts   | Operations     |
| Thomas    | Ziegler   | Operations     |
| Samantha  | Jameson   | Operations     |
| John      | Blake     | Operations     |
| Cindy     | Mason     | Operations     |
| Frank     | Portman   | Operations     |
| Theresa   | Markham   | Operations     |
| Beth      | Fowler    | Operations     |
| Rick      | Tulman    | Operations     |
+-----------+-----------+----------------+
18 rows in set (0.00 sec)
```

If you do not specify the type of join, then the server will do an inner join by default. As you will see later in the book, however, there are several types of joins, so you should get in the habit of specifying the exact type of join that you require.

If the names of the columns used to join the two tables are identical, which is true in the previous query, you can use the using subclause instead of the on subclause, as in:

```
mysql> SELECT e.fname, e.lname, d.name
    -> FROM employee e INNER JOIN department d
    ->    USING (dept_id);
+-----------+-----------+----------------+
| fname     | lname     | name           |
+-----------+-----------+----------------+
| Michael   | Smith     | Administration |
| Susan     | Barker    | Administration |
| Robert    | Tyler     | Administration |
| Susan     | Hawthorne | Operations     |
| John      | Gooding   | Loans          |
| Helen     | Fleming   | Operations     |
| Chris     | Tucker    | Operations     |
| Sarah     | Parker    | Operations     |
| Jane      | Grossman  | Operations     |
| Paula     | Roberts   | Operations     |
| Thomas    | Ziegler   | Operations     |
| Samantha  | Jameson   | Operations     |
| John      | Blake     | Operations     |
| Cindy     | Mason     | Operations     |
| Frank     | Portman   | Operations     |
| Theresa   | Markham   | Operations     |
| Beth      | Fowler    | Operations     |
```

```
| Rick      | Tulman     | Operations      |
+-----------+------------+-----------------+
18 rows in set (0.01 sec)
```

Since using is a shorthand notation that you can use in only a specific situation, I prefer always to use the on subclause to avoid confusion.

The ANSI Join Syntax

The notation used throughout this book for joining tables was introduced in the SQL92 version of the ANSI SQL standard. All the major databases (Oracle Database, Microsoft SQL Server, MySQL, IBM DB2 Universal Database, and Sybase Adaptive Server) have adopted the SQL92 join syntax. Because most of these servers have been around since before the release of the SQL92 specification, they all include an older join syntax as well. For example, all these servers would understand the following variation of the previous query:

```
mysql> SELECT e.fname, e.lname, d.name
    -> FROM employee e, department d
    -> WHERE e.dept_id = d.dept_id;
+-----------+------------+-----------------+
| fname     | lname      | name            |
+-----------+------------+-----------------+
| Michael   | Smith      | Administration  |
| Susan     | Barker     | Administration  |
| Robert    | Tyler      | Administration  |
| Susan     | Hawthorne  | Operations      |
| John      | Gooding    | Loans           |
| Helen     | Fleming    | Operations      |
| Chris     | Tucker     | Operations      |
| Sarah     | Parker     | Operations      |
| Jane      | Grossman   | Operations      |
| Paula     | Roberts    | Operations      |
| Thomas    | Ziegler    | Operations      |
| Samantha  | Jameson    | Operations      |
| John      | Blake      | Operations      |
| Cindy     | Mason      | Operations      |
| Frank     | Portman    | Operations      |
| Theresa   | Markham    | Operations      |
| Beth      | Fowler     | Operations      |
| Rick      | Tulman     | Operations      |
+-----------+------------+-----------------+
18 rows in set (0.01 sec)
```

This older method of specifying joins does not include the on subclause; instead, tables are named in the from clause separated by commas, and join conditions are included in the where clause. While you may decide to ignore the SQL92 syntax in favor of the older join syntax, the ANSI join syntax has the following advantages:

- Join conditions and filter conditions are separated into two different clauses (the on subclause and the where clause, respectively), making a query easier to understand.

- The join conditions for each pair of tables are contained in their own on clause, making it less likely that part of a join will be mistakenly omitted.

- Queries that use the SQL92 join syntax are portable across database servers, whereas the older syntax is slightly different across the different servers.

The benefits of the SQL92 join syntax are easier to identify for complex queries that include both join and filter conditions. Consider the following query, which returns all accounts opened by experienced tellers (hired prior to 2007) currently assigned to the Woburn branch:

```
mysql> SELECT a.account_id, a.cust_id, a.open_date, a.product_cd
    -> FROM account a, branch b, employee e
    -> WHERE a.open_emp_id = e.emp_id
    ->   AND e.start_date < '2007-01-01'
    ->   AND e.assigned_branch_id = b.branch_id
    ->   AND (e.title = 'Teller' OR e.title = 'Head Teller')
    ->   AND b.name = 'Woburn Branch';
+------------+---------+------------+------------+
| account_id | cust_id | open_date  | product_cd |
+------------+---------+------------+------------+
|          1 |       1 | 2000-01-15 | CHK        |
|          2 |       1 | 2000-01-15 | SAV        |
|          3 |       1 | 2004-06-30 | CD         |
|          4 |       2 | 2001-03-12 | CHK        |
|          5 |       2 | 2001-03-12 | SAV        |
|         17 |       7 | 2004-01-12 | CD         |
|         27 |      11 | 2004-03-22 | BUS        |
+------------+---------+------------+------------+
7 rows in set (0.00 sec)
```

With this query, it is not so easy to determine which conditions in the where clause are join conditions and which are filter conditions. It is also not readily apparent which type of join is being employed (to identify the type of join, you would need to look closely at the join conditions in the where clause to see whether any special characters are employed), nor is it easy to determine whether any join conditions have been mistakenly left out. Here's the same query using the SQL92 join syntax:

```
mysql> SELECT a.account_id, a.cust_id, a.open_date, a.product_cd
    -> FROM account a INNER JOIN employee e
    ->   ON a.open_emp_id = e.emp_id
    ->   INNER JOIN branch b
    ->   ON e.assigned_branch_id = b.branch_id
    -> WHERE e.start_date < '2007-01-01'
    ->   AND (e.title = 'Teller' OR e.title = 'Head Teller')
    ->   AND b.name = 'Woburn Branch';
+------------+---------+------------+------------+
| account_id | cust_id | open_date  | product_cd |
+------------+---------+------------+------------+
|          1 |       1 | 2000-01-15 | CHK        |
|          2 |       1 | 2000-01-15 | SAV        |
|          3 |       1 | 2004-06-30 | CD         |
|          4 |       2 | 2001-03-12 | CHK        |
|          5 |       2 | 2001-03-12 | SAV        |
```

```
|       17 |        7 | 2004-01-12 | CD  |
|       27 |       11 | 2004-03-22 | BUS |
+------------+----------+------------+----------+
7 rows in set (0.05 sec)
```

Hopefully, you will agree that the version using SQL92 join syntax is easier to understand.

Joining Three or More Tables

Joining three tables is similar to joining two tables, but with one slight wrinkle. With a two-table join, there are two tables and one join type in the from clause, and a single on subclause to define how the tables are joined. With a three-table join, there are three tables and two join types in the from clause, and two on subclauses. Here's another example of a query with a two-table join:

```
mysql> SELECT a.account_id, c.fed_id
    -> FROM account a INNER JOIN customer c
    ->   ON a.cust_id = c.cust_id
    -> WHERE c.cust_type_cd = 'B';
+------------+------------+
| account_id | fed_id     |
+------------+------------+
|         24 | 04-1111111 |
|         25 | 04-1111111 |
|         27 | 04-2222222 |
|         28 | 04-3333333 |
|         29 | 04-4444444 |
+------------+------------+
5 rows in set (0.15 sec)
```

This query, which returns the account ID and federal tax number for all business accounts, should look fairly straightforward by now. If, however, you add the employee table to the query to also retrieve the name of the teller who opened each account, it looks as follows:

```
mysql> SELECT a.account_id, c.fed_id, e.fname, e.lname
    -> FROM account a INNER JOIN customer c
    ->   ON a.cust_id = c.cust_id
    ->   INNER JOIN employee e
    ->   ON a.open_emp_id = e.emp_id
    -> WHERE c.cust_type_cd = 'B';
+------------+------------+---------+---------+
| account_id | fed_id     | fname   | lname   |
+------------+------------+---------+---------+
|         24 | 04-1111111 | Theresa | Markham |
|         25 | 04-1111111 | Theresa | Markham |
|         27 | 04-2222222 | Paula   | Roberts |
|         28 | 04-3333333 | Theresa | Markham |
|         29 | 04-4444444 | John    | Blake   |
+------------+------------+---------+---------+
5 rows in set (0.00 sec)
```

Now three tables, two join types, and two `on` subclauses are listed in the `from` clause, so things have gotten quite a bit busier. At first glance, the order in which the tables are named might cause you to think that the `employee` table is being joined to the `customer` table, since the `account` table is named first, followed by the `customer` table, and then the `employee` table. If you switch the order in which the first two tables appear, however, you will get the exact same results:

```
mysql> SELECT a.account_id, c.fed_id, e.fname, e.lname
    -> FROM customer c INNER JOIN account a
    ->   ON a.cust_id = c.cust_id
    ->   INNER JOIN employee e
    ->   ON a.open_emp_id = e.emp_id
    -> WHERE c.cust_type_cd = 'B';
+------------+------------+---------+---------+
| account_id | fed_id     | fname   | lname   |
+------------+------------+---------+---------+
|         24 | 04-1111111 | Theresa | Markham |
|         25 | 04-1111111 | Theresa | Markham |
|         27 | 04-2222222 | Paula   | Roberts |
|         28 | 04-3333333 | Theresa | Markham |
|         29 | 04-4444444 | John    | Blake   |
+------------+------------+---------+---------+
5 rows in set (0.09 sec)
```

The `customer` table is now listed first, followed by the `account` table and then the `employee` table. Since the on subclauses haven't changed, the results are the same. For the sake of completeness, here's the same query one last time, but with the table order completely reversed (`employee` to `account` to `customer`):

```
mysql> SELECT a.account_id, c.fed_id, e.fname, e.lname
    -> FROM employee e INNER JOIN account a
    ->   ON e.emp_id = a.open_emp_id
    ->   INNER JOIN customer c
    ->   ON a.cust_id = c.cust_id
    -> WHERE c.cust_type_cd = 'B';
+------------+------------+---------+---------+
| account_id | fed_id     | fname   | lname   |
+------------+------------+---------+---------+
|         24 | 04-1111111 | Theresa | Markham |
|         25 | 04-1111111 | Theresa | Markham |
|         27 | 04-2222222 | Paula   | Roberts |
|         28 | 04-3333333 | Theresa | Markham |
|         29 | 04-4444444 | John    | Blake   |
+------------+------------+---------+---------+
5 rows in set (0.00 sec)
```

Does Join Order Matter?

If you are confused about why all three versions of the `account`/`employee`/`customer` query yield the same results, keep in mind that SQL is a nonprocedural language, meaning that you describe what you want to retrieve and which database objects need to be involved, but it is up to the database server to determine how best to execute your query. Using statistics gathered from your database objects, the server must pick one of three tables as a starting point (the chosen table is thereafter known as the *driving table*), and then decide in which order to join the remaining tables. Therefore, the order in which tables appear in your `from` clause is not significant.

If, however, you believe that the tables in your query should always be joined in a particular order, you can place the tables in the desired order and then specify the keyword `STRAIGHT_JOIN` in MySQL, request the `FORCE ORDER` option in SQL Server, or use either the `ORDERED` or the `LEADING` optimizer hint in Oracle Database. For example, to tell the MySQL server to use the `customer` table as the driving table and to then join the `account` and `employee` tables, you could do the following:

```
mysql> SELECT STRAIGHT_JOIN a.account_id, c.fed_id, e.fname, e.lname
    -> FROM customer c INNER JOIN account a
    ->   ON a.cust_id = c.cust_id
    ->   INNER JOIN employee e
    ->   ON a.open_emp_id = e.emp_id
    -> WHERE c.cust_type_cd = 'B';
```

One way to think of a query that uses three or more tables is as a snowball rolling down a hill. The first two tables get the ball rolling, and each subsequent table gets tacked on to the snowball as it heads downhill. You can think of the snowball as the *intermediate result set*, which is picking up more and more columns as subsequent tables are joined. Therefore, the `employee` table is not really being joined to the `account` table, but rather the intermediate result set created when the `customer` and `account` tables were joined. (In case you were wondering why I chose a snowball analogy, I wrote this chapter in the midst of a New England winter: 110 inches so far, and more coming tomorrow. Oh joy.)

Using Subqueries As Tables

You have already seen several examples of queries that use three tables, but there is one variation worth mentioning: what to do if some of the data sets are generated by subqueries. Subqueries is the focus of Chapter 9, but I already introduced the concept of a subquery in the `from` clause in the previous chapter. Here's another version of an earlier query (find all accounts opened by experienced tellers currently assigned to the Woburn branch) that joins the `account` table to subqueries against the `branch` and `employee` tables:

```
1 SELECT a.account_id, a.cust_id, a.open_date, a.product_cd
2 FROM account a INNER JOIN
3   (SELECT emp_id, assigned_branch_id
```

```
 4     FROM employee
 5     WHERE start_date < '2007-01-01'
 6       AND (title = 'Teller' OR title = 'Head Teller')) e
 7   ON a.open_emp_id = e.emp_id
 8   INNER JOIN
 9    (SELECT branch_id
10     FROM branch
11     WHERE name = 'Woburn Branch') b
12   ON e.assigned_branch_id = b.branch_id;
```

The first subquery, which starts on line 3 and is given the alias e, finds all experienced tellers. The second subquery, which starts on line 9 and is given the alias b, finds the ID of the Woburn branch. First, the `account` table is joined to the experienced-teller subquery using the employee ID and then the table that results is joined to the Woburn branch subquery using the branch ID. The results are the same as those of the previous version of the query (try it and see for yourself), but the queries look very different from one another.

There isn't really anything shocking here, but it might take a minute to figure out what's going on. Notice, for example, the lack of a `where` clause in the main query; since all the filter conditions are against the `employee` and `branch` tables, the filter conditions are all inside the subqueries, so there is no need for any filter conditions in the main query. One way to visualize what is going on is to run the subqueries and look at the result sets. Here are the results of the first subquery against the `employee` table:

```
mysql> SELECT emp_id, assigned_branch_id
    -> FROM employee
    -> WHERE start_date < '2007-01-01'
    ->   AND (title = 'Teller' OR title = 'Head Teller');
+--------+--------------------+
| emp_id | assigned_branch_id |
+--------+--------------------+
|      8 |                  1 |
|      9 |                  1 |
|     10 |                  2 |
|     11 |                  2 |
|     13 |                  3 |
|     14 |                  3 |
|     16 |                  4 |
|     17 |                  4 |
|     18 |                  4 |
+--------+--------------------+
9 rows in set (0.03 sec)
```

Thus, this result set consists of a set of employee IDs and their corresponding branch IDs. When they are joined to the `account` table via the `emp_id` column, you now have an intermediate result set consisting of all rows from the `account` table with the additional column holding the branch ID of the employee that opened each account. Here are the results of the second subquery against the `branch` table:

```
mysql> SELECT branch_id
    -> FROM branch
    -> WHERE name = 'Woburn Branch';
```

```
+-----------+
| branch_id |
+-----------+
|         2 |
+-----------+
1 row in set (0.02 sec)
```

This query returns a single row containing a single column: the ID of the Woburn branch. This table is joined to the `assigned_branch_id` column of the intermediate result set, causing all accounts opened by non-Woburn-based employees to be filtered out of the final result set.

Using the Same Table Twice

If you are joining multiple tables, you might find that you need to join the same table more than once. In the sample database, for example, there are foreign keys to the `branch` table from both the `account` table (the branch at which the account was opened) and the `employee` table (the branch at which the employee works). If you want to include *both* branches in your result set, you can include the `branch` table twice in the `from` clause, joined once to the `employee` table and once to the `account` table. For this to work, you will need to give each instance of the `branch` table a different alias so that the server knows which one you are referring to in the various clauses, as in:

```
mysql> SELECT a.account_id, e.emp_id,
    ->   b_a.name open_branch, b_e.name emp_branch
    -> FROM account a INNER JOIN branch b_a
    ->   ON a.open_branch_id = b_a.branch_id
    ->   INNER JOIN employee e
    ->   ON a.open_emp_id = e.emp_id
    ->   INNER JOIN branch b_e
    ->   ON e.assigned_branch_id = b_e.branch_id
    -> WHERE a.product_cd = 'CHK';
+------------+--------+----------------+----------------+
| account_id | emp_id | open_branch    | emp_branch     |
+------------+--------+----------------+----------------+
|         10 |      1 | Headquarters   | Headquarters   |
|         14 |      1 | Headquarters   | Headquarters   |
|         21 |      1 | Headquarters   | Headquarters   |
|          1 |     10 | Woburn Branch  | Woburn Branch  |
|          4 |     10 | Woburn Branch  | Woburn Branch  |
|          7 |     13 | Quincy Branch  | Quincy Branch  |
|         13 |     16 | So. NH Branch  | So. NH Branch  |
|         18 |     16 | So. NH Branch  | So. NH Branch  |
|         24 |     16 | So. NH Branch  | So. NH Branch  |
|         28 |     16 | So. NH Branch  | So. NH Branch  |
+------------+--------+----------------+----------------+
10 rows in set (0.16 sec)
```

This query shows who opened each checking account, what branch it was opened at, and to which branch the employee who opened the account is currently assigned. The `branch` table is included twice, with aliases `b_a` and `b_e`. By assigning different aliases to each instance of the `branch` table, the server is able to understand which instance you are referring to: the one joined to the `account` table, or the one joined to the `employee` table. Therefore, this is one example of a query that *requires* the use of table aliases.

Self-Joins

Not only can you include the same table more than once in the same query, but you can actually join a table to itself. This might seem like a strange thing to do at first, but there are valid reasons for doing so. The `employee` table, for example, includes a *self-referencing foreign key*, which means that it includes a column (`superior_emp_id`) that points to the primary key within the same table. This column points to the employee's manager (unless the employee is the head honcho, in which case the column is `null`). Using a *self-join*, you can write a query that lists every employee's name along with the name of his or her manager:

```
mysql> SELECT e.fname, e.lname,
    ->   e_mgr.fname mgr_fname, e_mgr.lname mgr_lname
    -> FROM employee e INNER JOIN employee e_mgr
    ->   ON e.superior_emp_id = e_mgr.emp_id;
+----------+-----------+-----------+-----------+
| fname    | lname     | mgr_fname | mgr_lname |
+----------+-----------+-----------+-----------+
| Susan    | Barker    | Michael   | Smith     |
| Robert   | Tyler     | Michael   | Smith     |
| Susan    | Hawthorne | Robert    | Tyler     |
| John     | Gooding   | Susan     | Hawthorne |
| Helen    | Fleming   | Susan     | Hawthorne |
| Chris    | Tucker    | Helen     | Fleming   |
| Sarah    | Parker    | Helen     | Fleming   |
| Jane     | Grossman  | Helen     | Fleming   |
| Paula    | Roberts   | Susan     | Hawthorne |
| Thomas   | Ziegler   | Paula     | Roberts   |
| Samantha | Jameson   | Paula     | Roberts   |
| John     | Blake     | Susan     | Hawthorne |
| Cindy    | Mason     | John      | Blake     |
| Frank    | Portman   | John      | Blake     |
| Theresa  | Markham   | Susan     | Hawthorne |
| Beth     | Fowler    | Theresa   | Markham   |
| Rick     | Tulman    | Theresa   | Markham   |
+----------+-----------+-----------+-----------+
17 rows in set (0.00 sec)
```

This query includes two instances of the `employee` table: one to provide employee names (with the table alias e), and the other to provide manager names (with the table alias e_mgr). The `on` subclause uses these aliases to join the `employee` table to itself via the `superior_emp_id` foreign key. This is another example of a query for which table aliases are required; otherwise, the server wouldn't know whether you are referring to an employee or an employee's manager.

While there are 18 rows in the `employee` table, the query returned only 17 rows; the president of the bank, Michael Smith, has no superior (his `superior_emp_id` column is `null`), so the join failed for his row. To include Michael Smith in the result set, you would need to use an outer join, which we cover in Chapter 10.

Equi-Joins Versus Non-Equi-Joins

All of the multitable queries shown thus far have employed *equi-joins*, meaning that values from the two tables must match for the join to succeed. An equi-join always employs an equals sign, as in:

```
ON e.assigned_branch_id = b.branch_id
```

While the majority of your queries will employ equi-joins, you can also join your tables via ranges of values, which are referred to as *non-equi-joins*. Here's an example of a query that joins by a range of values:

```
SELECT e.emp_id, e.fname, e.lname, e.start_date
FROM employee e INNER JOIN product p
  ON e.start_date >= p.date_offered
  AND e.start_date <= p.date_retired
WHERE p.name = 'no-fee checking';
```

This query joins two tables that have no foreign key relationships. The intent is to find all employees who began working for the bank while the No-Fee Checking product was being offered. Thus, an employee's start date must be between the date the product was offered and the date the product was retired.

You may also find a need for a *self-non-equi-join*, meaning that a table is joined to itself using a non-equi-join. For example, let's say that the operations manager has decided to have a chess tournament for all bank tellers. You have been asked to create a list of all the pairings. You might try joining the `employee` table to itself for all tellers (`title = 'Teller'`) and return all rows where the `emp_id`s don't match (since a person can't play chess against himself):

```
mysql> SELECT e1.fname, e1.lname, 'VS' vs, e2.fname, e2.lname
    -> FROM employee e1 INNER JOIN employee e2
    ->   ON e1.emp_id != e2.emp_id
    -> WHERE e1.title = 'Teller' AND e2.title = 'Teller';
+----------+----------+----+----------+----------+
| fname    | lname    | vs | fname    | lname    |
+----------+----------+----+----------+----------+
| Sarah    | Parker   | VS | Chris    | Tucker   |
| Jane     | Grossman | VS | Chris    | Tucker   |
| Thomas   | Ziegler  | VS | Chris    | Tucker   |
| Samantha | Jameson  | VS | Chris    | Tucker   |
| Cindy    | Mason    | VS | Chris    | Tucker   |
| Frank    | Portman  | VS | Chris    | Tucker   |
| Beth     | Fowler   | VS | Chris    | Tucker   |
| Rick     | Tulman   | VS | Chris    | Tucker   |
| Chris    | Tucker   | VS | Sarah    | Parker   |
```

```
| Jane     | Grossman | VS | Sarah    | Parker   |
| Thomas   | Ziegler  | VS | Sarah    | Parker   |
| Samantha | Jameson  | VS | Sarah    | Parker   |
| Cindy    | Mason    | VS | Sarah    | Parker   |
| Frank    | Portman  | VS | Sarah    | Parker   |
| Beth     | Fowler   | VS | Sarah    | Parker   |
| Rick     | Tulman   | VS | Sarah    | Parker   |
...
| Chris    | Tucker   | VS | Rick     | Tulman   |
| Sarah    | Parker   | VS | Rick     | Tulman   |
| Jane     | Grossman | VS | Rick     | Tulman   |
| Thomas   | Ziegler  | VS | Rick     | Tulman   |
| Samantha | Jameson  | VS | Rick     | Tulman   |
| Cindy    | Mason    | VS | Rick     | Tulman   |
| Frank    | Portman  | VS | Rick     | Tulman   |
| Beth     | Fowler   | VS | Rick     | Tulman   |
+----------+----------+----+----------+----------+
72 rows in set (0.01 sec)
```

You're on the right track, but the problem here is that for each pairing (e.g., Sarah Parker versus Chris Tucker), there is also a reverse pairing (e.g., Chris Tucker versus Sarah Parker). One way to achieve the desired results is to use the join condition `e1.emp_id < e2.emp_id` so that each teller is paired only with those tellers having a higher employee ID (you can also use `e1.emp_id > e2.emp_id` if you wish):

```
mysql> SELECT e1.fname, e1.lname, 'VS' vs, e2.fname, e2.lname
    -> FROM employee e1 INNER JOIN employee e2
    ->   ON e1.emp_id < e2.emp_id
    -> WHERE e1.title = 'Teller' AND e2.title = 'Teller';
+----------+----------+----+----------+----------+
| fname    | lname    | vs | fname    | lname    |
+----------+----------+----+----------+----------+
| Chris    | Tucker   | VS | Sarah    | Parker   |
| Chris    | Tucker   | VS | Jane     | Grossman |
| Sarah    | Parker   | VS | Jane     | Grossman |
| Chris    | Tucker   | VS | Thomas   | Ziegler  |
| Sarah    | Parker   | VS | Thomas   | Ziegler  |
| Jane     | Grossman | VS | Thomas   | Ziegler  |
| Chris    | Tucker   | VS | Samantha | Jameson  |
| Sarah    | Parker   | VS | Samantha | Jameson  |
| Jane     | Grossman | VS | Samantha | Jameson  |
| Thomas   | Ziegler  | VS | Samantha | Jameson  |
| Chris    | Tucker   | VS | Cindy    | Mason    |
| Sarah    | Parker   | VS | Cindy    | Mason    |
| Jane     | Grossman | VS | Cindy    | Mason    |
| Thomas   | Ziegler  | VS | Cindy    | Mason    |
| Samantha | Jameson  | VS | Cindy    | Mason    |
| Chris    | Tucker   | VS | Frank    | Portman  |
| Sarah    | Parker   | VS | Frank    | Portman  |
| Jane     | Grossman | VS | Frank    | Portman  |
| Thomas   | Ziegler  | VS | Frank    | Portman  |
| Samantha | Jameson  | VS | Frank    | Portman  |
| Cindy    | Mason    | VS | Frank    | Portman  |
| Chris    | Tucker   | VS | Beth     | Fowler   |
| Sarah    | Parker   | VS | Beth     | Fowler   |
```

```
| Jane     | Grossman | VS | Beth | Fowler |
| Thomas   | Ziegler  | VS | Beth | Fowler |
| Samantha | Jameson  | VS | Beth | Fowler |
| Cindy    | Mason    | VS | Beth | Fowler |
| Frank    | Portman  | VS | Beth | Fowler |
| Chris    | Tucker   | VS | Rick | Tulman |
| Sarah    | Parker   | VS | Rick | Tulman |
| Jane     | Grossman | VS | Rick | Tulman |
| Thomas   | Ziegler  | VS | Rick | Tulman |
| Samantha | Jameson  | VS | Rick | Tulman |
| Cindy    | Mason    | VS | Rick | Tulman |
| Frank    | Portman  | VS | Rick | Tulman |
| Beth     | Fowler   | VS | Rick | Tulman |
+----------+----------+----+----------+----------+
36 rows in set (0.00 sec)
```

You now have a list of 36 pairings, which is the correct number when choosing pairs of 9 distinct things.

Join Conditions Versus Filter Conditions

You are now familiar with the concept that join conditions belong in the on subclause, while filter conditions belong in the where clause. However, SQL is flexible as to where you place your conditions, so you will need to take care when constructing your queries. For example, the following query joins two tables using a single join condition, and also includes a single filter condition in the where clause:

```
mysql> SELECT a.account_id, a.product_cd, c.fed_id
    -> FROM account a INNER JOIN customer c
    ->   ON a.cust_id = c.cust_id
    -> WHERE c.cust_type_cd = 'B';
+------------+------------+------------+
| account_id | product_cd | fed_id     |
+------------+------------+------------+
|         24 | CHK        | 04-1111111 |
|         25 | BUS        | 04-1111111 |
|         27 | BUS        | 04-2222222 |
|         28 | CHK        | 04-3333333 |
|         29 | SBL        | 04-4444444 |
+------------+------------+------------+
5 rows in set (0.01 sec)
```

That was pretty straightforward, but what happens if you mistakenly put the filter condition in the on subclause instead of in the where clause?

```
mysql> SELECT a.account_id, a.product_cd, c.fed_id
    -> FROM account a INNER JOIN customer c
    ->   ON a.cust_id = c.cust_id
    ->     AND c.cust_type_cd = 'B';
+------------+------------+------------+
| account_id | product_cd | fed_id     |
+------------+------------+------------+
|         24 | CHK        | 04-1111111 |
```

```
    |          25 | BUS         | 04-1111111 |
    |          27 | BUS         | 04-2222222 |
    |          28 | CHK         | 04-3333333 |
    |          29 | SBL         | 04-4444444 |
    +-------------+-------------+------------+
    5 rows in set (0.01 sec)
```

As you can see, the second version, which has *both* conditions in the on subclause and has no where clause, generates the same results. What if both conditions are placed in the where clause but the from clause still uses the ANSI join syntax?

```
mysql> SELECT a.account_id, a.product_cd, c.fed_id
    -> FROM account a INNER JOIN customer c
    -> WHERE a.cust_id = c.cust_id
    ->   AND c.cust_type_cd = 'B';
+-------------+-------------+------------+
| account_id  | product_cd  | fed_id     |
+-------------+-------------+------------+
|          24 | CHK         | 04-1111111 |
|          25 | BUS         | 04-1111111 |
|          27 | BUS         | 04-2222222 |
|          28 | CHK         | 04-3333333 |
|          29 | SBL         | 04-4444444 |
+-------------+-------------+------------+
5 rows in set (0.01 sec)
```

Once again, the MySQL server has generated the same result set. It will be up to you to put your conditions in the proper place so that your queries are easy to understand and maintain.

Test Your Knowledge

The following exercises are designed to test your understanding of inner joins. Please see Appendix C for the solutions to these exercises.

Exercise 5-1

Fill in the blanks (denoted by <#>) for the following query to obtain the results that follow:

```
mysql> SELECT e.emp_id, e.fname, e.lname, b.name
    -> FROM employee e INNER JOIN <1> b
    ->   ON e.assigned_branch_id = b.<2>;
+--------+----------+-----------+--------------+
| emp_id | fname    | lname     | name         |
+--------+----------+-----------+--------------+
|      1 | Michael  | Smith     | Headquarters |
|      2 | Susan    | Barker    | Headquarters |
|      3 | Robert   | Tyler     | Headquarters |
|      4 | Susan    | Hawthorne | Headquarters |
```

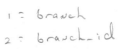

1 = branch
2 = branch_id

```
|      5 | John      | Gooding   | Headquarters   |
|      6 | Helen     | Fleming   | Headquarters   |
|      7 | Chris     | Tucker    | Headquarters   |
|      8 | Sarah     | Parker    | Headquarters   |
|      9 | Jane      | Grossman  | Headquarters   |
|     10 | Paula     | Roberts   | Woburn Branch  |
|     11 | Thomas    | Ziegler   | Woburn Branch  |
|     12 | Samantha  | Jameson   | Woburn Branch  |
|     13 | John      | Blake     | Quincy Branch  |
|     14 | Cindy     | Mason     | Quincy Branch  |
|     15 | Frank     | Portman   | Quincy Branch  |
|     16 | Theresa   | Markham   | So. NH Branch  |
|     17 | Beth      | Fowler    | So. NH Branch  |
|     18 | Rick      | Tulman    | So. NH Branch  |
+--------+-----------+-----------+----------------+
18 rows in set (0.03 sec)
```

Exercise 5-2

Write a query that returns the account ID for each nonbusiness customer (customer.cust_type_cd = 'I') with the customer's federal ID (customer.fed_id) and the name of the product on which the account is based (product.name).

SELECT a.account-id, c.fed-id, p.name
FROM account a INNER JOIN customer c ON a.cust-id = c.cust-id
INNER JOIN product p ON a.product-cd = p.product-cd
WHERE c.cust-type-cd = 'I';

Exercise 5-3

Construct a query that finds all employees whose supervisor is assigned to a different department. Retrieve the employees' ID, first name, and last name.

SELECT e.emp-id, e.fname, e.lname
FROM employee e INNER JOIN employee mgr ON e.superior.emp-id = mgr.emp-id
WHERE e.dept-id != mgr.dept-id');

CHAPTER 6

Working with Sets

Although you can interact with the data in a database one row at a time, relational databases are really all about sets. You have seen how you can create tables via queries or subqueries, make them persistent via `insert` statements, and bring them together via joins; this chapter explores how you can combine multiple tables using various set operators.

Set Theory Primer

In many parts of the world, basic set theory is included in elementary-level math curriculums. Perhaps you recall looking at something like what is shown in Figure 6-1.

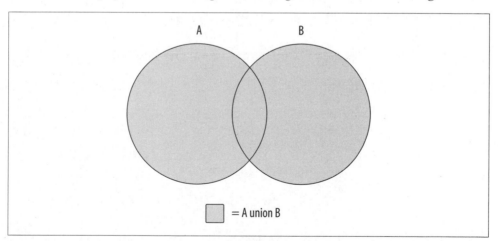

Figure 6-1. The union operation

The shaded area in Figure 6-1 represents the *union* of sets A and B, which is the combination of the two sets (with any overlapping regions included only once). Is this starting to look familiar? If so, then you'll finally get a chance to put that knowledge to use; if not, don't worry, because it's easy to visualize using a couple of diagrams.

Using circles to represent two data sets (A and B), imagine a subset of data that is common to both sets; this common data is represented by the overlapping area shown in Figure 6-1. Since set theory is rather uninteresting without an overlap between data sets, I use the same diagram to illustrate each set operation. There is another set operation that is concerned *only* with the overlap between two data sets; this operation is known as the *intersection* and is demonstrated in Figure 6-2.

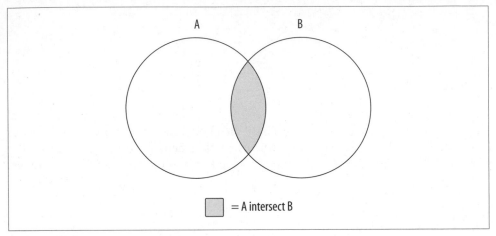

Figure 6-2. The intersection operation

The data set generated by the intersection of sets A and B is just the area of overlap between the two sets. If the two sets have no overlap, then the intersection operation yields the empty set.

The third and final set operation, which is demonstrated in Figure 6-3, is known as the *except* operation.

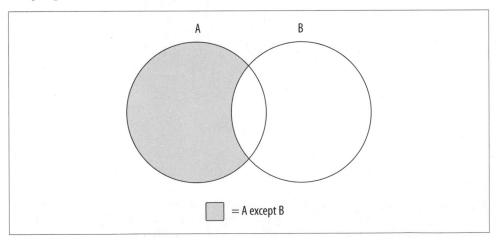

Figure 6-3. The except operation

Figure 6-3 shows the results of A except B, which is the whole of set A minus any overlap with set B. If the two sets have no overlap, then the operation A except B yields the whole of set A.

Using these three operations, or by combining different operations together, you can generate whatever results you need. For example, imagine that you want to build a set demonstrated by Figure 6-4.

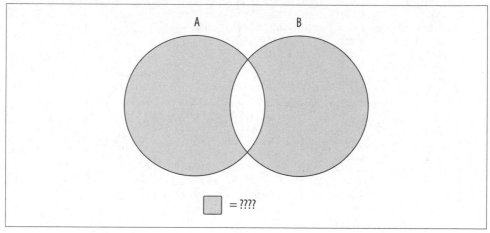

Figure 6-4. Mystery data set

The data set you are looking for includes all of sets A and B *without* the overlapping region. You can't achieve this outcome with just one of the three operations shown earlier; instead, you will need to first build a data set that encompasses all of sets A and B, and then utilize a second operation to remove the overlapping region. If the combined set is described as A union B, and the overlapping region is described as A intersect B, then the operation needed to generate the data set represented by Figure 6-4 would look as follows:

    (A union B) except (A intersect B)

Of course, there are often multiple ways to achieve the same results; you could reach a similar outcome using the following operation:

    (A except B) union (B except A)

While these concepts are fairly easy to understand using diagrams, the next sections show you how these concepts are applied to a relational database using the SQL set operators.

Set Theory in Practice

The circles used in the previous section's diagrams to represent data sets don't convey anything about what the data sets comprise. When dealing with actual data, however, there is a need to describe the composition of the data sets involved if they are to be combined. Imagine, for example, what would happen if you tried to generate the union of the `product` table and the `customer` table, whose table definitions are as follows:

```
mysql> DESC product;
+-----------------+-------------+------+-----+---------+-------+
| Field           | Type        | Null | Key | Default | Extra |
+-----------------+-------------+------+-----+---------+-------+
| product_cd      | varchar(10) | NO   | PRI | NULL    |       |
| name            | varchar(50) | NO   |     | NULL    |       |
| product_type_cd | varchar(10) | NO   | MUL | NULL    |       |
| date_offered    | date        | YES  |     | NULL    |       |
| date_retired    | date        | YES  |     | NULL    |       |
+-----------------+-------------+------+-----+---------+-------+
5 rows in set (0.23 sec)
mysql> DESC customer;
+--------------+------------------+------+-----+---------+----------------+
| Field        | Type             | Null | Key | Default | Extra          |
+--------------+------------------+------+-----+---------+----------------+
| cust_id      | int(10) unsigned | NO   | PRI | NULL    | auto_increment |
| fed_id       | varchar(12)      | NO   |     | NULL    |                |
| cust_type_cd | enum('I','B')    | NO   |     | NULL    |                |
| address      | varchar(30)      | YES  |     | NULL    |                |
| city         | varchar(20)      | YES  |     | NULL    |                |
| state        | varchar(20)      | YES  |     | NULL    |                |
| postal_code  | varchar(10)      | YES  |     | NULL    |                |
+--------------+------------------+------+-----+---------+----------------+
7 rows in set (0.04 sec)
```

When combined, the first column in the table that results would be the combination of the `product.product_cd` and `customer.cust_id` columns, the second column would be the combination of the `product.name` and `customer.fed_id` columns, and so forth. While some of the column pairs are easy to combine (e.g., two numeric columns), it is unclear how other column pairs should be combined, such as a numeric column with a string column or a string column with a date column. Additionally, the sixth and seventh columns of the combined tables would include data from only the `customer` table's sixth and seventh columns, since the `product` table has only five columns. Clearly, there needs to be some commonality between two tables that you wish to combine.

Therefore, when performing set operations on two data sets, the following guidelines must apply:

- Both data sets must have the same number of columns.
- The data types of each column across the two data sets must be the same (or the server must be able to convert one to the other).

With these rules in place, it is easier to envision what "overlapping data" means in practice; each column pair from the two sets being combined must contain the same string, number, or date for rows in the two tables to be considered the same.

You perform a set operation by placing a *set operator* between two `select` statements, as demonstrated by the following:

```
mysql> SELECT 1 num, 'abc' str
    -> UNION
    -> SELECT 9 num, 'xyz' str;
+-----+-----+
| num | str |
+-----+-----+
|   1 | abc |
|   9 | xyz |
+-----+-----+
2 rows in set (0.02 sec)
```

Each of the individual queries yields a data set consisting of a single row having a numeric column and a string column. The set operator, which in this case is `union`, tells the database server to combine all rows from the two sets. Thus, the final set includes two rows of two columns. This query is known as a *compound query* because it comprises multiple, otherwise-independent queries. As you will see later, compound queries may include *more* than two queries if multiple set operations are needed to attain the final results.

Set Operators

The SQL language includes three set operators that allow you to perform each of the various set operations described earlier in the chapter. Additionally, each set operator has two flavors, one that includes duplicates and another that removes duplicates (but not necessarily *all* of the duplicates). The following subsections define each operator and demonstrate how they are used.

The union Operator

The `union` and `union all` operators allow you to combine multiple data sets. The difference between the two is that `union` sorts the combined set and removes duplicates, whereas `union all` does not. With `union all`, the number of rows in the final data set will always equal the sum of the number of rows in the sets being combined. This operation is the simplest set operation to perform (from the server's point of view), since there is no need for the server to check for overlapping data. The following example demonstrates how you can use the `union all` operator to generate a full set of customer data from the two customer subtype tables:

```
mysql> SELECT 'IND' type_cd, cust_id, lname name
    -> FROM individual
    -> UNION ALL
    -> SELECT 'BUS' type_cd, cust_id, name
    -> FROM business;
+---------+---------+------------------------+
| type_cd | cust_id | name                   |
+---------+---------+------------------------+
```

```
| IND     |       1 | Hadley                 |
| IND     |       2 | Tingley                |
| IND     |       3 | Tucker                 |
| IND     |       4 | Hayward                |
| IND     |       5 | Frasier                |
| IND     |       6 | Spencer                |
| IND     |       7 | Young                  |
| IND     |       8 | Blake                  |
| IND     |       9 | Farley                 |
| BUS     |      10 | Chilton Engineering    |
| BUS     |      11 | Northeast Cooling Inc. |
| BUS     |      12 | Superior Auto Body     |
| BUS     |      13 | AAA Insurance Inc.     |
+---------+---------+------------------------+
13 rows in set (0.04 sec)
```

The query returns all 13 customers, with nine rows coming from the individual table and the other four coming from the business table. While the business table includes a single column to hold the company name, the individual table includes two name columns, one each for the person's first and last names. In this case, I chose to include only the last name from the individual table.

Just to drive home the point that the union all operator doesn't remove duplicates, here's the same query as the previous example but with an additional query against the business table:

```
mysql> SELECT 'IND' type_cd, cust_id, lname name
    -> FROM individual
    -> UNION ALL
    -> SELECT 'BUS' type_cd, cust_id, name
    -> FROM business
    -> UNION ALL
    -> SELECT 'BUS' type_cd, cust_id, name
    -> FROM business;
+---------+---------+------------------------+
| type_cd | cust_id | name                   |
+---------+---------+------------------------+
| IND     |       1 | Hadley                 |
| IND     |       2 | Tingley                |
| IND     |       3 | Tucker                 |
| IND     |       4 | Hayward                |
| IND     |       5 | Frasier                |
| IND     |       6 | Spencer                |
| IND     |       7 | Young                  |
| IND     |       8 | Blake                  |
| IND     |       9 | Farley                 |
| BUS     |      10 | Chilton Engineering    |
| BUS     |      11 | Northeast Cooling Inc. |
| BUS     |      12 | Superior Auto Body     |
| BUS     |      13 | AAA Insurance Inc.     |
| BUS     |      10 | Chilton Engineering    |
| BUS     |      11 | Northeast Cooling Inc. |
| BUS     |      12 | Superior Auto Body     |
| BUS     |      13 | AAA Insurance Inc.     |
```

```
+---------+---------+-----------------------+
```
17 rows in set (0.01 sec)

This compound query includes three `select` statements, two of which are identical. As you can see by the results, the four rows from the `business` table are included twice (customer IDs 10, 11, 12, and 13).

While you are unlikely to repeat the same query twice in a compound query, here is another compound query that returns duplicate data:

```
mysql> SELECT emp_id
    -> FROM employee
    -> WHERE assigned_branch_id = 2
    ->   AND (title = 'Teller' OR title = 'Head Teller')
    -> UNION ALL
    -> SELECT DISTINCT open_emp_id
    -> FROM account
    -> WHERE open_branch_id = 2;
+--------+
| emp_id |
+--------+
|     10 |
|     11 |
|     12 |
|     10 |
+--------+
4 rows in set (0.01 sec)
```

The first query in the compound statement retrieves all tellers assigned to the Woburn branch, whereas the second query returns the distinct set of tellers who opened accounts at the Woburn branch. Of the four rows in the result set, one of them is a duplicate (employee ID 10). If you would like your combined table to *exclude* duplicate rows, you need to use the `union` operator instead of `union all`:

```
mysql> SELECT emp_id
    -> FROM employee
    -> WHERE assigned_branch_id = 2
    ->   AND (title = 'Teller' OR title = 'Head Teller')
    -> UNION
    -> SELECT DISTINCT open_emp_id
    -> FROM account
    -> WHERE open_branch_id = 2;
+--------+
| emp_id |
+--------+
|     10 |
|     11 |
|     12 |
+--------+
3 rows in set (0.01 sec)
```

For this version of the query, only the three distinct rows are included in the result set, rather than the four rows (three distinct, one duplicate) returned when using `union all`.

The intersect Operator

The ANSI SQL specification includes the `intersect` operator for performing intersections. Unfortunately, version 6.0 of MySQL does not implement the `intersect` operator. If you are using Oracle or SQL Server 2008, you will be able to use `intersect`; since I am using MySQL for all examples in this book, however, the result sets for the example queries in this section are fabricated and cannot be executed with any versions up to and including version 6.0. I also refrain from showing the MySQL prompt (`mysql>`), since the statements are not being executed by the MySQL server.

If the two queries in a compound query return nonoverlapping data sets, then the intersection will be an empty set. Consider the following query:

```
SELECT emp_id, fname, lname
FROM employee
INTERSECT
SELECT cust_id, fname, lname
FROM individual;
Empty set (0.04 sec)
```

The first query returns the ID and name of each employee, while the second query returns the ID and name of each customer. These sets are completely nonoverlapping, so the intersection of the two sets yields the empty set.

The next step is to identify two queries that *do* have overlapping data and then apply the `intersect` operator. For this purpose, I use the same query used to demonstrate the difference between `union` and `union all`, except this time using `intersect`:

```
SELECT emp_id
FROM employee
WHERE assigned_branch_id = 2
  AND (title = 'Teller' OR title = 'Head Teller')
INTERSECT
SELECT DISTINCT open_emp_id
FROM account
WHERE open_branch_id = 2;
+--------+
| emp_id |
+--------+
|     10 |
+--------+
1 row in set (0.01 sec)
```

The intersection of these two queries yields employee ID 10, which is the only value found in both queries' result sets.

Along with the `intersect` operator, which removes any duplicate rows found in the overlapping region, the ANSI SQL specification calls for an `intersect all` operator, which does not remove duplicates. The only database server that currently implements the `intersect all` operator is IBM's DB2 Universal Server.

The except Operator

The ANSI SQL specification includes the except operator for performing the except operation. Once again, unfortunately, version 6.0 of MySQL does not implement the except operator, so the same rules apply for this section as for the previous section.

 If you are using Oracle Database, you will need to use the non-ANSI-compliant minus operator instead.

The except operator returns the first table minus any overlap with the second table. Here's the example from the previous section, but using except instead of intersect:

```
SELECT emp_id
FROM employee
WHERE assigned_branch_id = 2
  AND (title = 'Teller' OR title = 'Head Teller')
EXCEPT
SELECT DISTINCT open_emp_id
FROM account
WHERE open_branch_id = 2;
+--------+
| emp_id |
+--------+
|     11 |
|     12 |
+--------+
2 rows in set (0.01 sec)
```

In this version of the query, the result set consists of the three rows from the first query minus employee ID 10, which is found in the result sets from both queries. There is also an except all operator specified in the ANSI SQL specification, but once again, only IBM's DB2 Universal Server has implemented the except all operator.

The except all operator is a bit tricky, so here's an example to demonstrate how duplicate data is handled. Let's say you have two data sets that look as follows:

Set A

```
+--------+
| emp_id |
+--------+
|     10 |
|     11 |
|     12 |
|     10 |
|     10 |
+--------+
```

Set B

```
+--------+
| emp_id |
+--------+
|     10 |
|     10 |
+--------+
```

The operation A except B yields the following:

```
+--------+
| emp_id |
+--------+
|     11 |
|     12 |
+--------+
```

If you change the operation to A except all B, you will see the following:

```
+--------+
| emp_id |
+--------+
|     10 |
|     11 |
|     12 |
+--------+
```

Therefore, the difference between the two operations is that except removes all occurrences of duplicate data from set A, whereas except all only removes one occurrence of duplicate data from set A for every occurrence in set B.

Set Operation Rules

The following sections outline some rules that you must follow when working with compound queries.

Sorting Compound Query Results

If you want the results of your compound query to be sorted, you can add an order by clause after the last query. When specifying column names in the order by clause, you will need to choose from the column names in the first query of the compound query. Frequently, the column names are the same for both queries in a compound query, but this does not need to be the case, as demonstrated by the following:

```
mysql> SELECT emp_id, assigned_branch_id
    -> FROM employee
    -> WHERE title = 'Teller'
    -> UNION
    -> SELECT open_emp_id, open_branch_id
    -> FROM account
    -> WHERE product_cd = 'SAV'
    -> ORDER BY emp_id;
```

```
+--------+-------------------+
| emp_id | assigned_branch_id |
+--------+-------------------+
|      1 |                 1 |
|      7 |                 1 |
|      8 |                 1 |
|      9 |                 1 |
|     10 |                 2 |
|     11 |                 2 |
|     12 |                 2 |
|     14 |                 3 |
|     15 |                 3 |
|     16 |                 4 |
|     17 |                 4 |
|     18 |                 4 |
+--------+-------------------+
12 rows in set (0.04 sec)
```

The column names specified in the two queries are different in this example. If you specify a column name from the second query in your order by clause, you will see the following error:

```
mysql> SELECT emp_id, assigned_branch_id
    -> FROM employee
    -> WHERE title = 'Teller'
    -> UNION
    -> SELECT open_emp_id, open_branch_id
    -> FROM account
    -> WHERE product_cd = 'SAV'
    -> ORDER BY open_emp_id;
ERROR 1054 (42S22): Unknown column 'open_emp_id' in 'order clause'
```

I recommend giving the columns in both queries identical column aliases in order to avoid this issue.

Set Operation Precedence

If your compound query contains more than two queries using different set operators, you need to think about the order in which to place the queries in your compound statement to achieve the desired results. Consider the following three-query compound statement:

```
mysql> SELECT cust_id
    -> FROM account
    -> WHERE product_cd IN ('SAV', 'MM')
    -> UNION ALL
    -> SELECT a.cust_id
    -> FROM account a INNER JOIN branch b
    ->   ON a.open_branch_id = b.branch_id
    -> WHERE b.name = 'Woburn Branch'
    -> UNION
    -> SELECT cust_id
    -> FROM account
    -> WHERE avail_balance BETWEEN 500 AND 2500;
```

```
+---------+
| cust_id |
+---------+
|       1 |
|       2 |
|       3 |
|       4 |
|       8 |
|       9 |
|       7 |
|      11 |
|       5 |
+---------+
9 rows in set (0.00 sec)
```

This compound query includes three queries that return sets of nonunique customer IDs; the first and second queries are separated with the union all operator, while the second and third queries are separated with the union operator. While it might not seem to make much difference where the union and union all operators are placed, it does, in fact, make a difference. Here's the same compound query with the set operators reversed:

```
mysql> SELECT cust_id
    -> FROM account
    -> WHERE product_cd IN ('SAV', 'MM')
    -> UNION
    -> SELECT a.cust_id
    -> FROM account a INNER JOIN branch b
    ->   ON a.open_branch_id = b.branch_id
    -> WHERE b.name = 'Woburn Branch'
    -> UNION ALL
    -> SELECT cust_id
    -> FROM account
    -> WHERE avail_balance BETWEEN 500 AND 2500;
+---------+
| cust_id |
+---------+
|       1 |
|       2 |
|       3 |
|       4 |
|       8 |
|       9 |
|       7 |
|      11 |
|       1 |
|       1 |
|       2 |
|       3 |
|       3 |
|       4 |
|       4 |
|       5 |
|       9 |
```

```
+----------+
17 rows in set (0.00 sec)
```

Looking at the results, it's obvious that it *does* make a difference how the compound query is arranged when using different set operators. In general, compound queries containing three or more queries are evaluated in order from top to bottom, but with the following caveats:

- The ANSI SQL specification calls for the `intersect` operator to have precedence over the other set operators.

- You may dictate the order in which queries are combined by enclosing multiple queries in parentheses.

However, since MySQL does not yet implement `intersect` or allow parentheses in compound queries, you will need to carefully arrange the queries in your compound query so that you achieve the desired results. If you are using a different database server, you can wrap adjoining queries in parentheses to override the default top-to-bottom processing of compound queries, as in:

```
(SELECT cust_id
 FROM account
 WHERE product_cd IN ('SAV', 'MM')
 UNION ALL
 SELECT a.cust_id
 FROM account a INNER JOIN branch b
   ON a.open_branch_id = b.branch_id
 WHERE b.name = 'Woburn Branch')
INTERSECT
(SELECT cust_id
 FROM account
 WHERE avail_balance BETWEEN 500 AND 2500
 EXCEPT
 SELECT cust_id
 FROM account
 WHERE product_cd = 'CD'
  AND avail_balance < 1000);
```

For this compound query, the first and second queries would be combined using the `union all` operator, then the third and fourth queries would be combined using the `except` operator, and finally, the results from these two operations would be combined using the `intersect` operator to generate the final result set.

Test Your Knowledge

The following exercises are designed to test your understanding of set operations. See Appendix C for answers to these exercises.

Exercise 6-1

If set A = {L M N O P} and set B = {P Q R S T}, what sets are generated by the following operations?

- A union B   L M N O P Q R S T
- A union all B   L M N O P P Q R S T
- A intersect B   P
- A except B   L M N O P

Exercise 6-2

Write a compound query that finds the first and last names of all individual customers along with the first and last names of all employees.

```
SELECT i.fname, i.lname
FROM INDIVIDUAL
UNION
SELECT e.fname, e.lname
FROM employee
```

Exercise 6-3

Sort the results from Exercise 6-2 by the lname column.

```
ORDER BY lname
```

CHAPTER 7

Data Generation, Conversion, and Manipulation

As I mentioned in the Preface, this book strives to teach generic SQL techniques that can be applied across multiple database servers. This chapter, however, deals with the generation, conversion, and manipulation of string, numeric, and temporal data, and the SQL language does not include commands covering this functionality. Rather, built-in functions are used to facilitate data generation, conversion, and manipulation, and while the SQL standard does specify some functions, the database vendors often do not comply with the function specifications.

Therefore, my approach for this chapter is to show you some of the common ways in which data is manipulated within SQL statements, and then demonstrate some of the built-in functions implemented by Microsoft SQL Server, Oracle Database, and MySQL. Along with reading this chapter, I strongly recommend you purchase a reference guide covering all the functions implemented by your server. If you work with more than one database server, there are several reference guides that cover multiple servers, such as Kevin Kline et al.'s *SQL in a Nutshell* (*http://oreilly.com/catalog/9780596518844/*) and Jonathan Gennick's *SQL Pocket Guide* (*http://oreilly.com/catalog/9780596526887/*), both from O'Reilly.

Working with String Data

When working with string data, you will be using one of the following character data types:

CHAR

Holds fixed-length, blank-padded strings. MySQL allows CHAR values up to 255 characters in length, Oracle Database permits up to 2,000 characters, and SQL Server allows up to 8,000 characters.

varchar

> Holds variable-length strings. MySQL permits up to 65,535 characters in a `varchar` column, Oracle Database (via the `varchar2` type) allows up to 4,000 characters, and SQL Server allows up to 8,000 characters.

text *(MySQL and SQL Server)* or CLOB *(Character Large Object; Oracle Database)*

> Holds very large variable-length strings (generally referred to as documents in this context). MySQL has multiple text types (`tinytext`, `text`, `mediumtext`, and `long text`) for documents up to 4 GB in size. SQL Server has a single `text` type for documents up to 2 GB in size, and Oracle Database includes the `CLOB` data type, which can hold documents up to a whopping 128 TB. SQL Server 2005 also includes the `varchar(max)` data type and recommends its use instead of the `text` type, which will be removed from the server in some future release.

To demonstrate how you can use these various types, I use the following table for some of the examples in this section:

```
CREATE TABLE string_tbl
 (char_fld CHAR(30),
  vchar_fld VARCHAR(30),
  text_fld TEXT
 );
```

The next two subsections show how you can generate and manipulate string data.

String Generation

The simplest way to populate a character column is to enclose a string in quotes, as in:

```
mysql> INSERT INTO string_tbl (char_fld, vchar_fld, text_fld)
    -> VALUES ('This is char data',
    ->  'This is varchar data',
    ->  'This is text data');
Query OK, 1 row affected (0.00 sec)
```

When inserting string data into a table, remember that if the length of the string exceeds the maximum size for the character column (either the designated maximum or the maximum allowed for the data type), the server will throw an exception. Although this is the default behavior for all three servers, you can configure MySQL and SQL Server to silently truncate the string instead of throwing an exception. To demonstrate how MySQL handles this situation, the following update statement attempts to modify the `vchar_fld` column, whose maximum length is defined as 30, with a string that is 46 characters in length:

```
mysql> UPDATE string_tbl
    -> SET vchar_fld = 'This is a piece of extremely long varchar data';
ERROR 1406 (22001): Data too long for column 'vchar_fld' at row 1
```

With MySQL 6.0, the default behavior is now "strict" mode, which means that exceptions are thrown when problems arise, whereas in older versions of the server the string would have been truncated and a warning issued. If you would rather have the engine truncate the string and issue a warning instead of raising an exception, you can opt to be in ANSI mode. The following example shows how to check which mode you are in, and then how to change the mode using the SET command:

```
mysql> SELECT @@session.sql_mode;
+------------------------------------------------------------------+
| @@session.sql_mode                                               |
+------------------------------------------------------------------+
| STRICT_TRANS_TABLES,NO_AUTO_CREATE_USER,NO_ENGINE_SUBSTITUTION |
+------------------------------------------------------------------+
1 row in set (0.00 sec)

mysql> SET sql_mode='ansi';
Query OK, 0 rows affected (0.08 sec)

mysql> SELECT @@session.sql_mode;
+-----------------------------------------------------------------+
| @@session.sql_mode                                              |
+-----------------------------------------------------------------+
| REAL_AS_FLOAT,PIPES_AS_CONCAT,ANSI_QUOTES,IGNORE_SPACE,ANSI |
+-----------------------------------------------------------------+
1 row in set (0.00 sec)
```

If you rerun the previous UPDATE statement, you will find that the column has been modified, but the following warning is generated:

```
mysql> SHOW WARNINGS;
+---------+------+-------------------------------------------------+
| Level   | Code | Message                                         |
+---------+------+-------------------------------------------------+
| Warning | 1265 | Data truncated for column 'vchar_fld' at row 1 |
+---------+------+-------------------------------------------------+
1 row in set (0.00 sec)
```

If you retrieve the vchar_fld column, you will see that the string has indeed been truncated:

```
mysql> SELECT vchar_fld
    -> FROM string_tbl;
+------------------------------+
| vchar_fld                    |
+------------------------------+
| This is a piece of extremely |
+------------------------------+
1 row in set (0.05 sec)
```

As you can see, only the first 30 characters of the 46-character string made it into the vchar_fld column. The best way to avoid string truncation (or exceptions, in the case of Oracle Database or MySQL in strict mode) when working with varchar columns is to set the upper limit of a column to a high enough value to handle the longest strings that might be stored in the column (keeping in mind that the server allocates only enough space to store the string, so it is not wasteful to set a high upper limit for varchar columns).

Including single quotes

Since strings are demarcated by single quotes, you will need to be alert for strings that include single quotes or apostrophes. For example, you won't be able to insert the following string because the server will think that the apostrophe in the word *doesn't* marks the end of the string:

```
UPDATE string_tbl
SET text_fld = 'This string doesn't work';
```

To make the server ignore the apostrophe in the word *doesn't*, you will need to add an *escape* to the string so that the server treats the apostrophe like any other character in the string. All three servers allow you to escape a single quote by adding another single quote directly before, as in:

```
mysql> UPDATE string_tbl
    -> SET text_fld = 'This string didn''t work, but it does now';
Query OK, 1 row affected (0.01 sec)
Rows matched: 1  Changed: 1  Warnings: 0
```

 Oracle Database and MySQL users may also choose to escape a single quote by adding a backslash character immediately before, as in:

```
UPDATE string_tbl SET text_fld =
    'This string didn\'t work, but it does now'
```

If you retrieve a string for use in a screen or report field, you don't need to do anything special to handle embedded quotes:

```
mysql> SELECT text_fld
    -> FROM string_tbl;
+------------------------------------------+
| text_fld                                 |
+------------------------------------------+
| This string didn't work, but it does now |
+------------------------------------------+
1 row in set (0.00 sec)
```

However, if you are retrieving the string to add to a file that another program will read, you may want to include the escape as part of the retrieved string. If you are using MySQL, you can use the built-in function **quote()**, which places quotes around the entire string *and* adds escapes to any single quotes/apostrophes within the string. Here's what our string looks like when retrieved via the **quote()** function:

```
mysql> SELECT quote(text_fld)
    -> FROM string_tbl;
+-------------------------------------------+
| QUOTE(text_fld)                           |
+-------------------------------------------+
| 'This string didn\'t work, but it does now' |
+-------------------------------------------+
1 row in set (0.04 sec)
```

When retrieving data for data export, you may want to use the quote() function for all non-system-generated character columns, such as a customer_notes column.

Including special characters

If your application is multinational in scope, you might find yourself working with strings that include characters that do not appear on your keyboard. When working with the French and German languages, for example, you might need to include accented characters such as é and ö. The SQL Server and MySQL servers include the built-in function char() so that you can build strings from any of the 255 characters in the ASCII character set (Oracle Database users can use the chr() function). To demonstrate, the next example retrieves a typed string and its equivalent built via individual characters:

```
mysql> SELECT 'abcdefg', CHAR(97,98,99,100,101,102,103);
+---------+-------------------------------+
| abcdefg | CHAR(97,98,99,100,101,102,103) |
+---------+-------------------------------+
| abcdefg | abcdefg                       |
+---------+-------------------------------+
1 row in set (0.01 sec)
```

Thus, the 97th character in the ASCII character set is the letter *a*. While the characters shown in the preceding example are not special, the following examples show the location of the accented characters along with other special characters, such as currency symbols:

```
mysql> SELECT CHAR(128,129,130,131,132,133,134,135,136,137);
+-----------------------------------------------+
| CHAR(128,129,130,131,132,133,134,135,136,137) |
+-----------------------------------------------+
| Çüéâäàåçê ë                                    |
+-----------------------------------------------+
1 row in set (0.01 sec)

mysql> SELECT CHAR(138,139,140,141,142,143,144,145,146,147);
+-----------------------------------------------+
| CHAR(138,139,140,141,142,143,144,145,146,147) |
+-----------------------------------------------+
| èïîìÄÅÉæÆô                                     |
+-----------------------------------------------+
1 row in set (0.01 sec)

mysql> SELECT CHAR(148,149,150,151,152,153,154,155,156,157);
+-----------------------------------------------+
| CHAR(148,149,150,151,152,153,154,155,156,157) |
+-----------------------------------------------+
| öòÛùÿ...Ü¢£¥                                   |
+-----------------------------------------------+
1 row in set (0.00 sec)

mysql> SELECT CHAR(158,159,160,161,162,163,164,165);
+-------------------------------------+
```

```
| CHAR(158,159,160,161,162,163,164,165) |
+----------------------------------------+
| fáíóúñÑ                                |
+----------------------------------------+
1 row in set (0.01 sec)
```

I am using the latin1 character set for the examples in this section. If your session is configured for a different character set, you will see a different set of characters than what is shown here. The same concepts apply, but you will need to familiarize yourself with the layout of your character set to locate specific characters.

Building strings character by character can be quite tedious, especially if only a few of the characters in the string are accented. Fortunately, you can use the concat() function to concatenate individual strings, some of which you can type while others you can generate via the char() function. For example, the following shows how to build the phrase *danke schön* using the concat() and char() functions:

```
mysql> SELECT CONCAT('danke sch', CHAR(148), 'n');
+-------------------------------------+
| CONCAT('danke sch', CHAR(148), 'n') |
+-------------------------------------+
| danke schön                         |
+-------------------------------------+
1 row in set (0.00 sec)
```

Oracle Database users can use the concatenation operator (||) instead of the concat() function, as in:

```
SELECT 'danke sch' || CHR(148) || 'n'
FROM dual;
```

SQL Server does not include a concat() function, so you will need to use the concatenation operator (+), as in:

```
SELECT 'danke sch' + CHAR(148) + 'n'
```

If you have a character and need to find its ASCII equivalent, you can use the ascii() function, which takes the leftmost character in the string and returns a number:

```
mysql> SELECT ASCII('ö');
+------------+
| ASCII('ö') |
+------------+
|        148 |
+------------+
1 row in set (0.00 sec)
```

Using the char(), ascii(), and concat() functions (or concatenation operators), you should be able to work with any Roman language even if you are using a keyboard that does not include accented or special characters.

String Manipulation

Each database server includes many built-in functions for manipulating strings. This section explores two types of string functions: those that return numbers and those that return strings. Before I begin, however, I reset the data in the `string_tbl` table to the following:

```
mysql> DELETE FROM string_tbl;
Query OK, 1 row affected (0.02 sec)

mysql> INSERT INTO string_tbl (char_fld, vchar_fld, text_fld)
    -> VALUES ('This string is 28 characters',
    ->   'This string is 28 characters',
    ->   'This string is 28 characters');
Query OK, 1 row affected (0.00 sec)
```

String functions that return numbers

Of the string functions that return numbers, one of the most commonly used is the `length()` function, which returns the number of characters in the string (SQL Server users will need to use the `len()` function). The following query applies the `length()` function to each column in the `string_tbl` table:

```
mysql> SELECT LENGTH(char_fld) char_length,
    ->   LENGTH(vchar_fld) varchar_length,
    ->   LENGTH(text_fld) text_length
    -> FROM string_tbl;
+-------------+----------------+-------------+
| char_length | varchar_length | text_length |
+-------------+----------------+-------------+
|          28 |             28 |          28 |
+-------------+----------------+-------------+
1 row in set (0.00 sec)
```

While the lengths of the `varchar` and `text` columns are as expected, you might have expected the length of the `char` column to be 30, since I told you that strings stored in `char` columns are right-padded with spaces. The MySQL server removes trailing spaces from `char` data when it is retrieved, however, so you will see the same results from all string functions regardless of the type of column in which the strings are stored.

Along with finding the length of a string, you might want to find the location of a substring within a string. For example, if you want to find the position at which the string `'characters'` appears in the `vchar_fld` column, you could use the `position()` function, as demonstrated by the following:

```
mysql> SELECT POSITION('characters' IN vchar_fld)
    -> FROM string_tbl;
+-------------------------------------+
| POSITION('characters' IN vchar_fld) |
+-------------------------------------+
|                                  19 |
+-------------------------------------+
1 row in set (0.12 sec)
```

If the substring cannot be found, the position() function returns 0.

 For those of you who program in a language such as C or C++, where the first element of an array is at position 0, remember when working with databases that the first character in a string is at position 1. A return value of 0 from position() indicates that the substring could not be found, not that the substring was found at the first position in the string.

If you want to start your search at something other than the first character of your target string, you will need to use the locate() function, which is similar to the position() function except that it allows an optional third parameter, which is used to define the search's start position. The locate() function is also proprietary, whereas the position() function is part of the SQL:2003 standard. Here's an example asking for the position of the string 'is' starting at the fifth character in the vchar_fld column:

```
mysql> SELECT LOCATE('is', vchar_fld, 5)
    -> FROM string_tbl;
+----------------------------+
| LOCATE('is', vchar_fld, 5) |
+----------------------------+
|                         13 |
+----------------------------+
1 row in set (0.02 sec)
```

 Oracle Database does not include the position() or locate() function, but it does include the instr() function, which mimics the position() function when provided with two arguments and mimics the locate() function when provided with three arguments. SQL Server also doesn't include a position() or locate() function, but it does include the charindx() function, which also accepts either two or three arguments similar to Oracle's instr() function.

Another function that takes strings as arguments and returns numbers is the string comparison function strcmp(). Strcmp(), which is implemented only by MySQL and has no analog in Oracle Database or SQL Server, takes two strings as arguments, and returns one of the following:

- -1 if the first string comes before the second string in sort order
- 0 if the strings are identical
- 1 if the first string comes after the second string in sort order

To illustrate how the function works, I first show the sort order of five strings using a query, and then show how the strings compare to one another using strcmp(). Here are the five strings that I insert into the string_tbl table:

```
mysql> DELETE FROM string_tbl;
Query OK, 1 row affected (0.00 sec)
```

```
mysql> INSERT INTO string_tbl(vchar_fld) VALUES ('abcd');
Query OK, 1 row affected (0.03 sec)

mysql> INSERT INTO string_tbl(vchar_fld) VALUES ('xyz');
Query OK, 1 row affected (0.00 sec)

mysql> INSERT INTO string_tbl(vchar_fld) VALUES ('QRSTUV');
Query OK, 1 row affected (0.00 sec)

mysql> INSERT INTO string_tbl(vchar_fld) VALUES ('qrstuv');
Query OK, 1 row affected (0.00 sec)

mysql> INSERT INTO string_tbl(vchar_fld) VALUES ('12345');
Query OK, 1 row affected (0.00 sec)
```

Here are the five strings in their sort order:

```
mysql> SELECT vchar_fld
    -> FROM string_tbl
    -> ORDER BY vchar_fld;
+-----------+
| vchar_fld |
+-----------+
| 12345     |
| abcd      |
| QRSTUV    |
| qrstuv    |
| xyz       |
+-----------+
5 rows in set (0.00 sec)
```

The next query makes six comparisons among the five different strings:

```
mysql> SELECT STRCMP('12345','12345') 12345_12345,
    ->   STRCMP('abcd','xyz') abcd_xyz,
    ->   STRCMP('abcd','QRSTUV') abcd_QRSTUV,
    ->   STRCMP('qrstuv','QRSTUV') qrstuv_QRSTUV,
    ->   STRCMP('12345','xyz') 12345_xyz,
    ->   STRCMP('xyz','qrstuv') xyz_qrstuv;
+-------------+----------+-------------+---------------+-----------+------------+
| 12345_12345 | abcd_xyz | abcd_QRSTUV | qrstuv_QRSTUV | 12345_xyz | xyz_qrstuv |
+-------------+----------+-------------+---------------+-----------+------------+
|           0 |       -1 |          -1 |             0 |        -1 |          1 |
+-------------+----------+-------------+---------------+-----------+------------+
1 row in set (0.00 sec)
```

The first comparison yields 0, which is to be expected since I compared a string to itself. The fourth comparison also yields 0, which is a bit surprising, since the strings are composed of the same letters, with one string all uppercase and the other all lowercase. The reason for this result is that MySQL's strcmp() function is case-insensitive, which is something to remember when using the function. The other four comparisons yield either -1 or 1 depending on whether the first string comes before or after the second string in sort order. For example, strcmp('abcd','xyz') yields -1, since the string 'abcd' comes before the string 'xyz'.

Along with the `strcmp()` function, MySQL also allows you to use the `like` and `regexp` operators to compare strings in the `select` clause. Such comparisons will yield 1 (for `true`) or 0 (for `false`). Therefore, these operators allow you to build expressions that return a number, much like the functions described in this section. Here's an example using `like`:

```
mysql> SELECT name, name LIKE '%ns' ends_in_ns
    -> FROM department;
+----------------+------------+
| name           | ends_in_ns |
+----------------+------------+
| Operations     |          1 |
| Loans          |          1 |
| Administration |          0 |
+----------------+------------+
3 rows in set (0.25 sec)
```

This example retrieves all the department names, along with an expression that returns 1 if the department name ends in "ns" or 0 otherwise. If you want to perform more complex pattern matches, you can use the `regexp` operator, as demonstrated by the following:

```
mysql> SELECT cust_id, cust_type_cd, fed_id,
    ->   fed_id REGEXP '.{3}-.{2}-.{4}' is_ss_no_format
    -> FROM customer;
+---------+--------------+-------------+-----------------+
| cust_id | cust_type_cd | fed_id      | is_ss_no_format |
+---------+--------------+-------------+-----------------+
|       1 | I            | 111-11-1111 |               1 |
|       2 | I            | 222-22-2222 |               1 |
|       3 | I            | 333-33-3333 |               1 |
|       4 | I            | 444-44-4444 |               1 |
|       5 | I            | 555-55-5555 |               1 |
|       6 | I            | 666-66-6666 |               1 |
|       7 | I            | 777-77-7777 |               1 |
|       8 | I            | 888-88-8888 |               1 |
|       9 | I            | 999-99-9999 |               1 |
|      10 | B            | 04-1111111  |               0 |
|      11 | B            | 04-2222222  |               0 |
|      12 | B            | 04-3333333  |               0 |
|      13 | B            | 04-4444444  |               0 |
+---------+--------------+-------------+-----------------+
13 rows in set (0.00 sec)
```

The fourth column of this query returns 1 if the value stored in the `fed_id` column matches the format for a Social Security number.

 SQL Server and Oracle Database users can achieve similar results by building `case` expressions, which I describe in detail in Chapter 11.

String functions that return strings

In some cases, you will need to modify existing strings, either by extracting part of the string or by adding additional text to the string. Every database server includes multiple functions to help with these tasks. Before I begin, I once again reset the data in the string_tbl table:

```
mysql> DELETE FROM string_tbl;
Query OK, 5 rows affected (0.00 sec)

mysql> INSERT INTO string_tbl (text_fld)
    -> VALUES ('This string was 29 characters');
Query OK, 1 row affected (0.01 sec)
```

Earlier in the chapter, I demonstrated the use of the concat() function to help build words that include accented characters. The concat() function is useful in many other situations, including when you need to append additional characters to a stored string. For instance, the following example modifies the string stored in the text_fld column by tacking an additional phrase on the end:

```
mysql> UPDATE string_tbl
    -> SET text_fld = CONCAT(text_fld, ', but now it is longer');
Query OK, 1 row affected (0.03 sec)
Rows matched: 1  Changed: 1  Warnings: 0
```

The contents of the text_fld column are now as follows:

```
mysql> SELECT text_fld
    -> FROM string_tbl;
+-----------------------------------------------------+
| text_fld                                            |
+-----------------------------------------------------+
| This string was 29 characters, but now it is longer |
+-----------------------------------------------------+
1 row in set (0.00 sec)
```

Thus, like all functions that return a string, you can use concat() to replace the data stored in a character column.

Another common use for the concat() function is to build a string from individual pieces of data. For example, the following query generates a narrative string for each bank teller:

```
mysql> SELECT CONCAT(fname, ' ', lname, ' has been a ',
    ->   title, ' since ', start_date) emp_narrative
    -> FROM employee
    -> WHERE title = 'Teller' OR title = 'Head Teller';
+-----------------------------------------------------+
| emp_narrative                                       |
+-----------------------------------------------------+
| Helen Fleming has been a Head Teller since 2008-03-17 |
| Chris Tucker has been a Teller since 2008-09-15     |
| Sarah Parker has been a Teller since 2006-12-02     |
| Jane Grossman has been a Teller since 2006-05-03    |
| Paula Roberts has been a Head Teller since 2006-07-27 |
```

```
| Thomas Ziegler has been a Teller since 2004-10-23     |
| Samantha Jameson has been a Teller since 2007-01-08   |
| John Blake has been a Head Teller since 2004-05-11    |
| Cindy Mason has been a Teller since 2006-08-09        |
| Frank Portman has been a Teller since 2007-04-01      |
| Theresa Markham has been a Head Teller since 2005-03-15 |
| Beth Fowler has been a Teller since 2006-06-29        |
| Rick Tulman has been a Teller since 2006-12-12        |
+-------------------------------------------------------+
13 rows in set (0.30 sec)
```

The concat() function can handle any expression that returns a string, and will even convert numbers and dates to string format, as evidenced by the date column (start_date) used as an argument. Although Oracle Database includes the concat() function, it will accept only two string arguments, so the previous query will not work on Oracle. Instead, you would need to use the concatenation operator (||) rather than a function call, as in:

```
SELECT fname || ' ' || lname || ' has been a ' ||
  title || ' since ' || start_date emp_narrative
FROM employee
WHERE title = 'Teller' OR title = 'Head Teller';
```

SQL Server does not include a concat() function, so you would need to use the same approach as the previous query, except that you would use SQL Server's concatenation operator (+) instead of ||.

While concat() is useful for adding characters to the beginning or end of a string, you may also have a need to add or replace characters in the *middle* of a string. All three database servers provide functions for this purpose, but all of them are different, so I demonstrate the MySQL function and then show the functions from the other two servers.

MySQL includes the insert() function, which takes four arguments: the original string, the position at which to start, the number of characters to replace, and the replacement string. Depending on the value of the third argument, the function may be used to either insert or replace characters in a string. With a value of 0 for the third argument, the replacement string is inserted and any trailing characters are pushed to the right, as in:

```
mysql> SELECT INSERT('goodbye world', 9, 0, 'cruel ') string;
+---------------------+
| string              |
+---------------------+
| goodbye cruel world |
+---------------------+
1 row in set (0.00 sec)
```

In this example, all characters starting from position 9 are pushed to the right and the string 'cruel' is inserted. If the third argument is greater than zero, then that number of characters is replaced with the replacement string, as in:

```
mysql> SELECT INSERT('goodbye world', 1, 7, 'hello') string;
+-------------+
| string      |
+-------------+
| hello world |
+-------------+
1 row in set (0.00 sec)
```

For this example, the first seven characters are replaced with the string 'hello'. Oracle Database does not provide a single function with the flexibility of MySQL's insert() function, but Oracle does provide the replace() function, which is useful for replacing one substring with another. Here's the previous example reworked to use replace():

```
SELECT REPLACE('goodbye world', 'goodbye', 'hello')
FROM dual;
```

All instances of the string 'goodbye' will be replaced with the string 'hello', resulting in the string 'hello world'. The replace() function will replace *every* instance of the search string with the replacement string, so you need to be careful that you don't end up with more replacements than you anticipated.

SQL Server also includes a replace() function with the same functionality as Oracle's, but SQL Server also includes a function called stuff() with similar functionality to MySQL's insert() function. Here's an example:

```
SELECT STUFF('hello world', 1, 5, 'goodbye cruel')
```

When executed, five characters are removed starting at position 1, and then the string 'goodbye cruel' is inserted at the starting position, resulting in the string 'goodbye cruel world'.

Along with inserting characters into a string, you may have a need to *extract* a substring from a string. For this purpose, all three servers include the substring() function (although Oracle Database's version is called substr()), which extracts a specified number of characters starting at a specified position. The following example extracts five characters from a string starting at the ninth position:

```
mysql> SELECT SUBSTRING('goodbye cruel world', 9, 5);
+---------------------------------------+
| SUBSTRING('goodbye cruel world', 9, 5) |
+---------------------------------------+
| cruel                                 |
+---------------------------------------+
1 row in set (0.00 sec)
```

Along with the functions demonstrated here, all three servers include many more built-in functions for manipulating string data. While many of them are designed for very specific purposes, such as generating the string equivalent of octal or hexadecimal numbers, there are many other general-purpose functions as well, such as functions that remove or add trailing spaces. For more information, consult your server's SQL reference guide, or a general-purpose SQL reference guide such as *SQL in a Nutshell* (O'Reilly).

Working with Numeric Data

Unlike string data (and temporal data, as you will see shortly), numeric data generation is quite straightforward. You can type a number, retrieve it from another column, or generate it via a calculation. All the usual arithmetic operators (+, -, *, /) are available for performing calculations, and parentheses may be used to dictate precedence, as in:

```
mysql> SELECT (37 * 59) / (78 - (8 * 6));
+----------------------------+
| (37 * 59) / (78 - (8 * 6)) |
+----------------------------+
|                      72.77 |
+----------------------------+
1 row in set (0.00 sec)
```

As I mentioned in Chapter 2, the main concern when storing numeric data is that numbers might be rounded if they are larger than the specified size for a numeric column. For example, the number 9.96 will be rounded to 10.0 if stored in a column defined as float(3,1).

Performing Arithmetic Functions

Most of the built-in numeric functions are used for specific arithmetic purposes, such as determining the square root of a number. Table 7-1 lists some of the common numeric functions that take a single numeric argument and return a number.

Table 7-1. Single-argument numeric functions

Function name	Description
Acos(x)	Calculates the arc cosine of x
Asin(x)	Calculates the arc sine of x
Atan(x)	Calculates the arc tangent of x
Cos(x)	Calculates the cosine of x
Cot(x)	Calculates the cotangent of x
Exp(x)	Calculates e^x
Ln(x)	Calculates the natural log of x
Sin(x)	Calculates the sine of x
Sqrt(x)	Calculates the square root of x
Tan(x)	Calculates the tangent of x

These functions perform very specific tasks, and I refrain from showing examples for these functions (if you don't recognize a function by name or description, then you probably don't need it). Other numeric functions used for calculations, however, are a bit more flexible and deserve some explanation.

For example, the `modulo` operator, which calculates the remainder when one number is divided into another number, is implemented in MySQL and Oracle Database via the `mod()` function. The following example calculates the remainder when 4 is divided into 10:

```
mysql> SELECT MOD(10,4);
+-----------+
| MOD(10,4) |
+-----------+
|         2 |
+-----------+
1 row in set (0.02 sec)
```

While the `mod()` function is typically used with integer arguments, with MySQL you can also use real numbers, as in:

```
mysql> SELECT MOD(22.75, 5);
+---------------+
| MOD(22.75, 5) |
+---------------+
|          2.75 |
+---------------+
1 row in set (0.02 sec)
```

 SQL Server does not have a `mod()` function. Instead, the operator `%` is used for finding remainders. The expression `10 % 4` will therefore yield the value `2`.

Another numeric function that takes two numeric arguments is the `pow()` function (or `power()` if you are using Oracle Database or SQL Server), which returns one number raised to the power of a second number, as in:

```
mysql> SELECT POW(2,8);
+----------+
| POW(2,8) |
+----------+
|      256 |
+----------+
1 row in set (0.03 sec)
```

Thus, `pow(2,8)` is the MySQL equivalent of specifying 2^8. Since computer memory is allocated in chunks of 2^x bytes, the `pow()` function can be a handy way to determine the exact number of bytes in a certain amount of memory:

```
mysql> SELECT POW(2,10) kilobyte, POW(2,20) megabyte,
    ->    POW(2,30) gigabyte, POW(2,40) terabyte;
+----------+----------+------------+----------------+
| kilobyte | megabyte | gigabyte   | terabyte       |
+----------+----------+------------+----------------+
|     1024 |  1048576 | 1073741824 | 1099511627776  |
+----------+----------+------------+----------------+
1 row in set (0.00 sec)
```

I don't know about you, but I find it easier to remember that a gigabyte is 2^{30} bytes than to remember the number 1,073,741,824.

Controlling Number Precision

When working with floating-point numbers, you may not always want to interact with or display a number with its full precision. For example, you may store monetary transaction data with a precision to six decimal places, but you might want to round to the nearest hundredth for display purposes. Four functions are useful when limiting the precision of floating-point numbers: ceil(), floor(), round(), and truncate(). All three servers include these functions, although Oracle Database includes trunc() instead of truncate(), and SQL Server includes ceiling() instead of ceil().

The ceil() and floor() functions are used to round either up or down to the closest integer, as demonstrated by the following:

```
mysql> SELECT CEIL(72.445), FLOOR(72.445);
+--------------+---------------+
| CEIL(72.445) | FLOOR(72.445) |
+--------------+---------------+
|           73 |            72 |
+--------------+---------------+
1 row in set (0.06 sec)
```

Thus, any number between 72 and 73 will be evaluated as 73 by the ceil() function and 72 by the floor() function. Remember that ceil() will round up even if the decimal portion of a number is very small, and floor() will round down even if the decimal portion is quite significant, as in:

```
mysql> SELECT CEIL(72.000000001), FLOOR(72.999999999);
+--------------------+---------------------+
| CEIL(72.000000001) | FLOOR(72.999999999) |
+--------------------+---------------------+
|                 73 |                  72 |
+--------------------+---------------------+
1 row in set (0.00 sec)
```

If this is a bit too severe for your application, you can use the round() function to round up or down from the *midpoint* between two integers, as in:

```
mysql> SELECT ROUND(72.49999), ROUND(72.5), ROUND(72.50001);
+-----------------+-------------+-----------------+
| ROUND(72.49999) | ROUND(72.5) | ROUND(72.50001) |
+-----------------+-------------+-----------------+
|              72 |          73 |              73 |
+-----------------+-------------+-----------------+
1 row in set (0.00 sec)
```

Using round(), any number whose decimal portion is halfway or more between two integers will be rounded up, whereas the number will be rounded down if the decimal portion is anything less than halfway between the two integers.

Most of the time, you will want to keep at least some part of the decimal portion of a number rather than rounding to the nearest integer; the round() function allows an optional second argument to specify how many digits to the right of the decimal place to round to. The next example shows how you can use the second argument to round the number 72.0909 to one, two, and three decimal places:

```
mysql> SELECT ROUND(72.0909, 1), ROUND(72.0909, 2), ROUND(72.0909, 3);
+-------------------+-------------------+-------------------+
| ROUND(72.0909, 1) | ROUND(72.0909, 2) | ROUND(72.0909, 3) |
+-------------------+-------------------+-------------------+
|              72.1 |             72.09 |            72.091 |
+-------------------+-------------------+-------------------+
1 row in set (0.00 sec)
```

Like the round() function, the truncate() function allows an optional second argument to specify the number of digits to the right of the decimal, but truncate() simply discards the unwanted digits without rounding. The next example shows how the number 72.0909 would be truncated to one, two, and three decimal places:

```
mysql> SELECT TRUNCATE(72.0909, 1), TRUNCATE(72.0909, 2),
    ->     TRUNCATE(72.0909, 3);
+----------------------+----------------------+----------------------+
| TRUNCATE(72.0909, 1) | TRUNCATE(72.0909, 2) | TRUNCATE(72.0909, 3) |
+----------------------+----------------------+----------------------+
|                 72.0 |                72.09 |               72.090 |
+----------------------+----------------------+----------------------+
1 row in set (0.00 sec)
```

 SQL Server does not include a truncate() function. Instead, the round() function allows for an optional third argument which, if present and nonzero, calls for the number to be truncated rather than rounded.

Both truncate() and round() also allow a *negative* value for the second argument, meaning that numbers to the *left* of the decimal place are truncated or rounded. This might seem like a strange thing to do at first, but there are valid applications. For example, you might sell a product that can be purchased only in units of 10. If a customer were to order 17 units, you could choose from one of the following methods to modify the customer's order quantity:

```
mysql> SELECT ROUND(17, -1), TRUNCATE(17, -1);
+---------------+------------------+
| ROUND(17, -1) | TRUNCATE(17, -1) |
+---------------+------------------+
|            20 |               10 |
+---------------+------------------+
1 row in set (0.00 sec)
```

If the product in question is thumbtacks, then it might not make much difference to your bottom line whether you sold the customer 10 or 20 thumbtacks when only 17 were requested; if you are selling Rolex watches, however, your business may fare better by rounding.

Handling Signed Data

If you are working with numeric columns that allow negative values (in Chapter 2, I showed how a numeric column may be labeled *unsigned*, meaning that only positive numbers are allowed), several numeric functions might be of use. Let's say, for example, that you are asked to generate a report showing the current status of each bank account. The following query returns three columns useful for generating the report:

```
mysql> SELECT account_id, SIGN(avail_balance), ABS(avail_balance)
    -> FROM account;
+------------+---------------------+--------------------+
| account_id | SIGN(avail_balance) | ABS(avail_balance) |
+------------+---------------------+--------------------+
|          1 |                   1 |            1057.75 |
|          2 |                   1 |             500.00 |
|          3 |                   1 |            3000.00 |
|          4 |                   1 |            2258.02 |
|          5 |                   1 |             200.00 |
| ...                                                   |
|         19 |                   1 |            1500.00 |
|         20 |                   1 |           23575.12 |
|         21 |                   0 |               0.00 |
|         22 |                   1 |            9345.55 |
|         23 |                   1 |           38552.05 |
|         24 |                   1 |           50000.00 |
+------------+---------------------+--------------------+
24 rows in set (0.00 sec)
```

The second column uses the `sign()` function to return -1 if the account balance is negative, 0 if the account balance is zero, and 1 if the account balance is positive. The third column returns the absolute value of the account balance via the `abs()` function.

Working with Temporal Data

Of the three types of data discussed in this chapter (character, numeric, and temporal), temporal data is the most involved when it comes to data generation and manipulation. Some of the complexity of temporal data is caused by the myriad ways in which a single date and time can be described. For example, the date on which I wrote this paragraph can be described in all the following ways:

- Wednesday, September 17, 2008
- 9/17/2008 2:14:56 P.M. EST
- 9/17/2008 19:14:56 GMT
- 2612008 (Julian format)
- Star date [-4] 85712.03 14:14:56 (*Star Trek* format)

While some of these differences are purely a matter of formatting, most of the complexity has to do with your frame of reference, which we explore in the next section.

Dealing with Time Zones

Because people around the world prefer that noon coincides roughly with the sun's peak at their location, there has never been a serious attempt to coerce everyone to use a universal clock. Instead, the world has been sliced into 24 imaginary sections, called *time zones*; within a particular time zone, everyone agrees on the current time, whereas people in different time zones do not. While this seems simple enough, some geographic regions shift their time by one hour twice a year (implementing what is known as *daylight saving time*) and some do not, so the time difference between two points on Earth might be four hours for one half of the year and five hours for the other half of the year. Even within a single time zone, different regions may or may not adhere to daylight saving time, causing different clocks in the same time zone to agree for one half of the year but be one hour different for the rest of the year.

While the computer age has exacerbated the issue, people have been dealing with time zone differences since the early days of naval exploration. To ensure a common point of reference for timekeeping, fifteenth-century navigators set their clocks to the time of day in Greenwich, England. This became known as *Greenwich Mean Time*, or GMT. All other time zones can be described by the number of hours' difference from GMT; for example, the time zone for the Eastern United States, known as *Eastern Standard Time*, can be described as GMT –5:00, or five hours earlier than GMT.

Today, we use a variation of GMT called *Coordinated Universal Time*, or UTC, which is based on an atomic clock (or, to be more precise, the average time of 200 atomic clocks in 50 locations worldwide, which is referred to as *Universal Time*). Both SQL Server and MySQL provide functions that will return the current UTC timestamp (getutcdate() for SQL Server and utc_timestamp() for MySQL).

Most database servers default to the time zone setting of the server on which it resides and provide tools for modifying the time zone if needed. For example, a database used to store stock exchange transactions from around the world would generally be configured to use UTC time, whereas a database used to store transactions at a particular retail establishment might use the server's time zone.

MySQL keeps two different time zone settings: a global time zone, and a session time zone, which may be different for each user logged in to a database. You can see both settings via the following query:

```
mysql> SELECT @@global.time_zone, @@session.time_zone;
+--------------------+--------------------+
| @@global.time_zone | @@session.time_zone |
+--------------------+--------------------+
| SYSTEM             | SYSTEM             |
+--------------------+--------------------+
1 row in set (0.00 sec)
```

A value of system tells you that the server is using the time zone setting from the server on which the database resides.

If you are sitting at a computer in Zurich, Switzerland, and you open a session across the network to a MySQL server situated in New York, you may want to change the time zone setting for your session, which you can do via the following command:

```
mysql> SET time_zone = 'Europe/Zurich';
Query OK, 0 rows affected (0.18 sec)
```

If you check the time zone settings again, you will see the following:

```
mysql> SELECT @@global.time_zone, @@session.time_zone;
+--------------------+---------------------+
| @@global.time_zone | @@session.time_zone |
+--------------------+---------------------+
| SYSTEM             | Europe/Zurich       |
+--------------------+---------------------+
1 row in set (0.00 sec)
```

All dates displayed in your session will now conform to Zurich time.

 Oracle Database users can change the time zone setting for a session via the following command:

```
ALTER SESSION TIMEZONE = 'Europe/Zurich'
```

Generating Temporal Data

You can generate temporal data via any of the following means:

- Copying data from an existing date, datetime, or time column
- Executing a built-in function that returns a date, datetime, or time
- Building a string representation of the temporal data to be evaluated by the server

To use the last method, you will need to understand the various components used in formatting dates.

String representations of temporal data

Table 2-5 in Chapter 2 presented the more popular date components; to refresh your memory, Table 7-2 shows these same components.

Loading MySQL Time Zone Data

If you are running the MySQL server on a Windows platform, you will need to load time zone data manually before you can set global or session time zones. To do so, you need to follow these steps:

---

1. Download the time zone data from *http://dev.mysql.com/downloads/timezones.html*.

2. Shut down your MySQL server.

3. Extract the files from the downloaded ZIP file (in my case, the file was called *timezone-2006p.zip*) and place them in your MySQL installation directory under */data/mysql* (the full path for my installation was */Program Files/MySQL/MySQL Server 6.0/data/mysql*).

4. Restart your MySQL server.

To look at the time zone data, change to the *mysql* database via the `use mysql` command, and execute the following query:

```
mysql> SELECT name FROM time_zone_name;
+--------------------------------+
| name                           |
+--------------------------------+
| Africa/Abidjan                 |
| Africa/Accra                   |
| Africa/Addis_Ababa             |
| Africa/Algiers                 |
| Africa/Asmera                  |
| Africa/Bamako                  |
| Africa/Bangui                  |
| Africa/Banjul                  |
| Africa/Bissau                  |
| Africa/Blantyre                |
| Africa/Brazzaville             |
| Africa/Bujumbura               |
...
| US/Alaska                      |
| US/Aleutian                    |
| US/Arizona                     |
| US/Central                     |
| US/East-Indiana                |
| US/Eastern                     |
| US/Hawaii                      |
| US/Indiana-Starke              |
| US/Michigan                    |
| US/Mountain                    |
| US/Pacific                     |
| US/Pacific-New                 |
| US/Samoa                       |
| UTC                            |
| W-SU                           |
| WET                            |
| Zulu                           |
+--------------------------------+
546 rows in set (0.01 sec)
```

To change your time zone setting, choose one of the names from the previous query that best matches your location.

Table 7-2. Date format components

Component	Definition	Range
YYYY	Year, including century	1000 to 9999
MM	Month	01 (January) to 12 (December)
DD	Day	01 to 31
HH	Hour	00 to 23
HHH	Hours (elapsed)	−838 to 838
MI	Minute	00 to 59
SS	Second	00 to 59

To build a string that the server can interpret as a `date`, `datetime`, or `time`, you need to put the various components together in the order shown in Table 7-3.

Table 7-3. Required date components

Type	Default format
Date	YYYY-MM-DD
Datetime	YYYY-MM-DD HH:MI:SS
Timestamp	YYYY-MM-DD HH:MI:SS
Time	HHH:MI:SS

Thus, to populate a `datetime` column with 3:30 P.M. on September 17, 2008, you will need to build the following string:

```
'2008-09-17 15:30:00'
```

If the server is expecting a `datetime` value, such as when updating a `datetime` column or when calling a built-in function that takes a `datetime` argument, you can provide a properly formatted string with the required date components, and the server will do the conversion for you. For example, here's a statement used to modify the date of a bank transaction:

```
UPDATE transaction
SET txn_date = '2008-09-17 15:30:00'
WHERE txn_id = 99999;
```

The server determines that the string provided in the `set` clause must be a `datetime` value, since the string is being used to populate a `datetime` column. Therefore, the server will attempt to convert the string for you by parsing the string into the six components (year, month, day, hour, minute, second) included in the default `datetime` format.

String-to-date conversions

If the server is *not* expecting a `datetime` value, or if you would like to represent the `datetime` using a nondefault format, you will need to tell the server to convert the string

---

to a `datetime`. For example, here is a simple query that returns a `datetime` value using the `cast()` function:

```
mysql> SELECT CAST('2008-09-17 15:30:00' AS DATETIME);
+-----------------------------------------+
| CAST('2008-09-17 15:30:00' AS DATETIME) |
+-----------------------------------------+
| 2008-09-17 15:30:00                     |
+-----------------------------------------+
1 row in set (0.00 sec)
```

We cover the `cast()` function at the end of this chapter. While this example demonstrates how to build `datetime` values, the same logic applies to the `date` and `time` types as well. The following query uses the `cast()` function to generate a `date` value and a `time` value:

```
mysql> SELECT CAST('2008-09-17' AS DATE) date_field,
    ->    CAST('108:17:57' AS TIME) time_field;
+------------+------------+
| date_field | time_field |
+------------+------------+
| 2008-09-17 | 108:17:57  |
+------------+------------+
1 row in set (0.00 sec)
```

You may, of course, explicitly convert your strings even when the server is expecting a `date`, `datetime`, or `time` value, rather than letting the server do an implicit conversion.

When strings are converted to temporal values—whether explicitly or implicitly—you must provide all the date components in the required order. While some servers are quite strict regarding the date format, the MySQL server is quite lenient about the separators used between the components. For example, MySQL will accept all of the following strings as valid representations of 3:30 P.M. on September 17, 2008:

```
'2008-09-17 15:30:00'
'2008/09/17 15:30:00'
'2008,09,17,15,30,00'
'20080917153000'
```

Although this gives you a bit more flexibility, you may find yourself trying to generate a temporal value *without* the default date components; the next section demonstrates a built-in function that is far more flexible than the `cast()` function.

Functions for generating dates

If you need to generate temporal data from a string, and the string is not in the proper form to use the `cast()` function, you can use a built-in function that allows you to provide a format string along with the date string. MySQL includes the `str_to_date()` function for this purpose. Say, for example, that you pull the string `'September 17, 2008'` from a file and need to use it to update a `date` column. Since the string is not in the required YYYY-MM-DD format, you can use `str_to_date()` instead of reformatting the string so that you can use the `cast()` function, as in:

```
UPDATE individual
SET birth_date = STR_TO_DATE('September 17, 2008', '%M %d, %Y')
WHERE cust_id = 9999;
```

The second argument in the call to str_to_date() defines the format of the date string, with, in this case, a month name (%M), a numeric day (%d), and a four-digit numeric year (%Y). While there are over 30 recognized format components, Table 7-4 defines the dozen or so most commonly used components.

Table 7-4. Date format components

Format component	Description
%M	Month name (January to December)
%m	Month numeric (01 to 12)
%d	Day numeric (01 to 31)
%j	Day of year (001 to 366)
%W	Weekday name (Sunday to Saturday)
%Y	Year, four-digit numeric
%y	Year, two-digit numeric
%H	Hour (00 to 23)
%h	Hour (01 to 12)
%i	Minutes (00 to 59)
%s	Seconds (00 to 59)
%f	Microseconds (000000 to 999999)
%p	A.M. or P.M.

The str_to_date() function returns a datetime, date, or time value depending on the contents of the format string. For example, if the format string includes only %H, %i, and %s, then a time value will be returned.

Oracle Database users can use the to_date() function in the same manner as MySQL's str_to_date() function. SQL Server includes a convert() function that is not quite as flexible as MySQL and Oracle Database; rather than supplying a custom format string, your date string must conform to one of 21 predefined formats.

If you are trying to generate the *current* date/time, then you won't need to build a string, because the following built-in functions will access the system clock and return the current date and/or time as a string for you:

```
mysql> SELECT CURRENT_DATE(), CURRENT_TIME(), CURRENT_TIMESTAMP();
+----------------+----------------+---------------------+
| CURRENT_DATE() | CURRENT_TIME() | CURRENT_TIMESTAMP() |
+----------------+----------------+---------------------+
| 2008-09-18     | 19:53:12       | 2008-09-18 19:53:12 |
+----------------+----------------+---------------------+
1 row in set (0.12 sec)
```

The values returned by these functions are in the default format for the temporal type being returned. Oracle Database includes current_date() and current_timestamp() but not current_time(), and SQL Server includes only the current_timestamp() function.

Manipulating Temporal Data

This section explores the built-in functions that take date arguments and return dates, strings, or numbers.

Temporal functions that return dates

Many of the built-in temporal functions take one date as an argument and return another date. MySQL's date_add() function, for example, allows you to add any kind of interval (e.g., days, months, years) to a specified date to generate another date. Here's an example that demonstrates how to add five days to the current date:

```
mysql> SELECT DATE_ADD(CURRENT_DATE(), INTERVAL 5 DAY);
+------------------------------------------+
| DATE_ADD(CURRENT_DATE(), INTERVAL 5 DAY) |
+------------------------------------------+
| 2008-09-22                               |
+------------------------------------------+
1 row in set (0.06 sec)
```

The second argument is composed of three elements: the interval keyword, the desired quantity, and the type of interval. Table 7-5 shows some of the commonly used interval types.

Table 7-5. Common interval types

Interval name	Description
Second	Number of seconds
Minute	Number of minutes
Hour	Number of hours
Day	Number of days
Month	Number of months
Year	Number of years
Minute_second	Number of minutes and seconds, separated by ":"
Hour_second	Number of hours, minutes, and seconds, separated by ":"
Year_month	Number of years and months, separated by "-"

While the first six types listed in Table 7-5 are pretty straightforward, the last three types require a bit more explanation since they have multiple elements. For example, if you are told that transaction ID 9999 actually occurred 3 hours, 27 minutes, and 11 seconds later than what was posted to the `transaction` table, you can fix it via the following:

```
UPDATE transaction
SET txn_date = DATE_ADD(txn_date, INTERVAL '3:27:11' HOUR_SECOND)
WHERE txn_id = 9999;
```

In this example, the `date_add()` function takes the value in the `txn_date` column, adds 3 hours, 27 minutes, and 11 seconds to it, and uses the value that results to modify the `txn_date` column.

Or, if you work in HR and found out that employee ID 4789 claimed to be younger than he actually is, you could add 9 years and 11 months to his birth date, as in:

```
UPDATE employee
SET birth_date = DATE_ADD(birth_date, INTERVAL '9-11' YEAR_MONTH)
WHERE emp_id = 4789;
```

 SQL Server users can accomplish the previous example using the `dateadd()` function:

```
UPDATE employee
SET birth_date =
  DATEADD(MONTH, 119, birth_date)
WHERE emp_id = 4789
```

SQL Server doesn't have combined intervals (i.e., `year_month`), so I converted 9 years, 11 months to 119 months.

Oracle Database users can use the `add_months()` function for this example, as in:

```
UPDATE employee
SET birth_date = ADD_MONTHS(birth_date, 119)
WHERE emp_id = 4789;
```

There are some cases where you want to add an interval to a date, and you know where you want to arrive but not how many days it takes to get there. For example, let's say that a bank customer logs on to the online banking system and schedules a transfer for the end of the month. Rather than writing some code that figures out what month you are currently in and looks up the number of days in that month, you can call the `last_day()` function, which does the work for you (both MySQL and Oracle Database include the `last_day()` function; SQL Server has no comparable function). If the customer asks for the transfer on September 17, 2008, you could find the last day of September via the following:

```
mysql> SELECT LAST_DAY('2008-09-17');
+------------------------+
| LAST_DAY('2008-09-17') |
+------------------------+
| 2008-09-30             |
+------------------------+
1 row in set (0.10 sec)
```

Whether you provide a date or datetime value, the last_day() function always returns a date. Although this function may not seem like an enormous timesaver, the underlying logic can be tricky if you're trying to find the last day of February and need to figure out whether the current year is a leap year.

Another temporal function that returns a date is one that converts a datetime value from one time zone to another. For this purpose, MySQL includes the convert_tz() function and Oracle Database includes the new_time() function. If I want to convert my current local time to UTC, for example, I could do the following:

```
mysql> SELECT CURRENT_TIMESTAMP() current_est,
    ->   CONVERT_TZ(CURRENT_TIMESTAMP(), 'US/Eastern', 'UTC') current_utc;
+---------------------+---------------------+
| current_est         | current_utc         |
+---------------------+---------------------+
| 2008-09-18 20:01:25 | 2008-09-19 00:01:25 |
+---------------------+---------------------+
1 row in set (0.76 sec)
```

This function comes in handy when receiving dates in a different time zone than what is stored in your database.

Temporal functions that return strings

Most of the temporal functions that return string values are used to extract a portion of a date or time. For example, MySQL includes the dayname() function to determine which day of the week a certain date falls on, as in:

```
mysql> SELECT DAYNAME('2008-09-18');
+-----------------------+
| DAYNAME('2008-09-18') |
+-----------------------+
| Thursday              |
+-----------------------+
1 row in set (0.08 sec)
```

Many such functions are included with MySQL for extracting information from date values, but I recommend that you use the extract() function instead, since it's easier to remember a few variations of one function than to remember a dozen different functions. Additionally, the extract() function is part of the SQL:2003 standard and has been implemented by Oracle Database as well as MySQL.

The extract() function uses the same interval types as the date_add() function (see Table 7-5) to define which element of the date interests you. For example, if you want to extract just the year portion of a datetime value, you can do the following:

```
mysql> SELECT EXTRACT(YEAR FROM '2008-09-18 22:19:05');
+-------------------------------------------+
| EXTRACT(YEAR FROM '2008-09-18 22:19:05') |
+-------------------------------------------+
|                                      2008 |
+-------------------------------------------+
1 row in set (0.00 sec)
```

 SQL Server doesn't include an implementation of extract(), but it does include the datepart() function. Here's how you would extract the year from a datetime value using datepart():

```
SELECT DATEPART(YEAR, GETDATE())
```

Temporal functions that return numbers

Earlier in this chapter, I showed you a function used to add a given interval to a date value, thus generating another date value. Another common activity when working with dates is to take *two* date values and determine the number of intervals (days, weeks, years) *between* the two dates. For this purpose, MySQL includes the function datediff(), which returns the number of full days between two dates. For example, if I want to know the number of days that my kids will be out of school this summer, I can do the following:

```
mysql> SELECT DATEDIFF('2009-09-03', '2009-06-24');
+--------------------------------------+
| DATEDIFF('2009-09-03', '2009-06-24') |
+--------------------------------------+
|                                   71 |
+--------------------------------------+
1 row in set (0.05 sec)
```

Thus, I will have to endure 71 days of poison ivy, mosquito bites, and scraped knees before the kids are safely back at school. The datediff() function ignores the time of day in its arguments. Even if I include a time-of-day, setting it to one second until midnight for the first date and to one second after midnight for the second date, those times will have no effect on the calculation:

```
mysql> SELECT DATEDIFF('2009-09-03 23:59:59', '2009-06-24 00:00:01');
+--------------------------------------------------------+
| DATEDIFF('2009-09-03 23:59:59', '2009-06-24 00:00:01') |
+--------------------------------------------------------+
|                                                     71 |
+--------------------------------------------------------+
1 row in set (0.00 sec)
```

If I switch the arguments and have the earlier date first, `datediff()` will return a negative number, as in:

```
mysql> SELECT DATEDIFF('2009-06-24', '2009-09-03');
+--------------------------------------+
| DATEDIFF('2009-06-24', '2009-09-03') |
+--------------------------------------+
|                                  -71 |
+--------------------------------------+
1 row in set (0.01 sec)
```

 SQL Server also includes the `datediff()` function, but it is more flexible than the MySQL implementation in that you can specify the interval type (i.e., year, month, day, hour) instead of counting only the number of days between two dates. Here's how SQL Server would accomplish the previous example:

```
SELECT DATEDIFF(DAY, '2009-06-24', '2009-09-03')
```

Oracle Database allows you to determine the number of days between two dates simply by subtracting one date from another.

Conversion Functions

Earlier in this chapter, I showed you how to use the `cast()` function to convert a string to a `datetime` value. While every database server includes a number of proprietary functions used to convert data from one type to another, I recommend using the `cast()` function, which is included in the SQL:2003 standard and has been implemented by MySQL, Oracle Database, and Microsoft SQL Server.

To use `cast()`, you provide a value or expression, the `as` keyword, and the type to which you want the value converted. Here's an example that converts a string to an integer:

```
mysql> SELECT CAST('1456328' AS SIGNED INTEGER);
+-----------------------------------+
| CAST('1456328' AS SIGNED INTEGER) |
+-----------------------------------+
|                           1456328 |
+-----------------------------------+
1 row in set (0.01 sec)
```

When converting a string to a number, the `cast()` function will attempt to convert the entire string from left to right; if any non-numeric characters are found in the string, the conversion halts without an error. Consider the following example:

```
mysql> SELECT CAST('999ABC111' AS UNSIGNED INTEGER);
+---------------------------------------+
| CAST('999ABC111' AS UNSIGNED INTEGER) |
+---------------------------------------+
|                                   999 |
+---------------------------------------+
1 row in set, 1 warning (0.08 sec)
```

```
mysql> show warnings;
+---------+------+---------------------------------------------------+
| Level   | Code | Message                                           |
+---------+------+---------------------------------------------------+
| Warning | 1292 | Truncated incorrect INTEGER value: '999ABC111'    |
+---------+------+---------------------------------------------------+
1 row in set (0.07 sec)
```

In this case, the first three digits of the string are converted, whereas the rest of the string is discarded, resulting in a value of 999. The server did, however, issue a warning to let you know that not all the string was converted.

If you are converting a string to a date, time, or datetime value, then you will need to stick with the default formats for each type, since you can't provide the cast() function with a format string. If your date string is not in the default format (i.e., YYYY-MM-DD HH:MI:SS for datetime types), then you will need to resort to using another function, such as MySQL's str_to_date() function described earlier in the chapter.

Test Your Knowledge

These exercises are designed to test your understanding of some of the built-in functions shown in this chapter. See Appendix C for the answers.

Exercise 7-1

Write a query that returns the 17th through 25th characters of the string 'Please find the substring in this string'.

Exercise 7-2

Write a query that returns the absolute value and sign (-1, 0, or 1) of the number -25.76823. Also return the number rounded to the nearest hundredth.

Exercise 7-3

Write a query to return just the month portion of the current date.

CHAPTER 8

Grouping and Aggregates

Data is generally stored at the lowest level of granularity needed by any of a database's users; if Chuck in accounting needs to look at individual customer transactions, then there needs to be a table in the database that stores individual transactions. That doesn't mean, however, that all users must deal with the data as it is stored in the database. The focus of this chapter is on how data can be grouped and aggregated to allow users to interact with it at some higher level of granularity than what is stored in the database.

Grouping Concepts

Sometimes you will want to find trends in your data that will require the database server to cook the data a bit before you can generate the results you are looking for. For example, let's say that you are in charge of operations at the bank, and you would like to find out how many accounts are being opened by each bank teller. You could issue a simple query to look at the raw data:

```
mysql> SELECT open_emp_id
    -> FROM account;
+-------------+
| open_emp_id |
+-------------+
|           1 |
|           1 |
|           1 |
|           1 |
|           1 |
|           1 |
|           1 |
|           1 |
|          10 |
|          10 |
|          10 |
|          10 |
|          10 |
|          10 |
|          10 |
|          13 |
```

```
|          13 |
|          13 |
|          16 |
|          16 |
|          16 |
|          16 |
|          16 |
|          16 |
+-------------+
24 rows in set (0.01 sec)
```

With only 24 rows in the account table, it is relatively easy to see that four different employees opened accounts and that employee ID 16 has opened six accounts; however, if the bank has dozens of employees and thousands of accounts, this approach would prove tedious and error-prone.

Instead, you can ask the database server to group the data for you by using the group by clause. Here's the same query but employing a group by clause to group the account data by employee ID:

```
mysql> SELECT open_emp_id
    -> FROM account
    -> GROUP BY open_emp_id;
+-------------+
| open_emp_id |
+-------------+
|           1 |
|          10 |
|          13 |
|          16 |
+-------------+
4 rows in set (0.00 sec)
```

The result set contains one row for each distinct value in the open_emp_id column, resulting in four rows instead of the full 24 rows. The reason for the smaller result set is that each of the four employees opened more than one account. To see how many accounts each teller opened, you can use an *aggregate function* in the select clause to count the number of rows in each group:

```
mysql> SELECT open_emp_id, COUNT(*) how_many
    -> FROM account
    -> GROUP BY open_emp_id;
+-------------+----------+
| open_emp_id | how_many |
+-------------+----------+
|           1 |        8 |
|          10 |        7 |
|          13 |        3 |
|          16 |        6 |
+-------------+----------+
4 rows in set (0.00 sec)
```

The aggregate function count() counts the number of rows in each group, and the asterisk tells the server to count everything in the group. Using the combination of a

---

group by clause and the count() aggregate function, you are able to generate exactly the data needed to answer the business question without having to look at the raw data.

When grouping data, you may need to filter out undesired data from your result set based on groups of data rather than based on the raw data. Since the group by clause runs *after* the where clause has been evaluated, you cannot add filter conditions to your where clause for this purpose. For example, here's an attempt to filter out any cases where an employee has opened fewer than five accounts:

```
mysql> SELECT open_emp_id, COUNT(*) how_many
    -> FROM account
    -> WHERE COUNT(*) > 4
    -> GROUP BY open_emp_id;
ERROR 1111 (HY000): Invalid use of group function
```

You cannot refer to the aggregate function count(*) in your where clause, because the groups have not yet been generated at the time the where clause is evaluated. Instead, you must put your group filter conditions in the having clause. Here's what the query would look like using having:

```
mysql> SELECT open_emp_id, COUNT(*) how_many
    -> FROM account
    -> GROUP BY open_emp_id
    -> HAVING COUNT(*) > 4;
+-------------+----------+
| open_emp_id | how_many |
+-------------+----------+
|           1 |        8 |
|          10 |        7 |
|          16 |        6 |
+-------------+----------+
3 rows in set (0.00 sec)
```

Because those groups containing fewer than five members have been filtered out via the having clause, the result set now contains only those employees who have opened five or more accounts, thus eliminating employee ID 13 from the results.

Aggregate Functions

Aggregate functions perform a specific operation over all rows in a group. Although every database server has its own set of specialty aggregate functions, the common aggregate functions implemented by all major servers include:

Max()
    Returns the maximum value within a set

Min()
    Returns the minimum value within a set

Avg()
    Returns the average value across a set

Sum( )

 Returns the sum of the values across a set

Count( )

 Returns the number of values in a set

Here's a query that uses all of the common aggregate functions to analyze the available balances for all checking accounts:

```
mysql> SELECT MAX(avail_balance) max_balance,
    ->   MIN(avail_balance) min_balance,
    ->   AVG(avail_balance) avg_balance,
    ->   SUM(avail_balance) tot_balance,
    ->   COUNT(*) num_accounts
    -> FROM account
    -> WHERE product_cd = 'CHK';
+-------------+-------------+-------------+-------------+--------------+
| max_balance | min_balance | avg_balance | tot_balance | num_accounts |
+-------------+-------------+-------------+-------------+--------------+
|    38552.05 |      122.37 | 7300.800985 |    73008.01 |           10 |
+-------------+-------------+-------------+-------------+--------------+
1 row in set (0.09 sec)
```

The results from this query tell you that, across the 10 checking accounts in the `account` table, there is a maximum balance of $38,552.05, a minimum balance of $122.37, an average balance of $7,300.80, and a total balance across all 10 accounts of $73,008.01. Hopefully, this gives you an appreciation for the role of these aggregate functions; the next subsections further clarify how you can utilize these functions.

Implicit Versus Explicit Groups

In the previous example, every value returned by the query is generated by an aggregate function, and the aggregate functions are applied across the group of rows specified by the filter condition `product_cd = 'CHK'`. Since there is no `group by` clause, there is a single, *implicit* group (all rows returned by the query).

In most cases, however, you will want to retrieve additional columns along with columns generated by aggregate functions. What if, for example, you wanted to extend the previous query to execute the same five aggregate functions for *each* product type, instead of just for checking accounts? For this query, you would want to retrieve the `product_cd` column along with the five aggregate functions, as in:

```
SELECT product_cd,
    MAX(avail_balance) max_balance,
    MIN(avail_balance) min_balance,
    AVG(avail_balance) avg_balance,
    SUM(avail_balance) tot_balance,
    COUNT(*) num_accounts
FROM account;
```

However, if you try to execute the query, you will receive the following error:

```
ERROR 1140 (42000): Mixing of GROUP columns (MIN(),MAX(),COUNT(),...) with no GROUP
columns is illegal if there is no GROUP BY clause
```

While it may be obvious to you that you want the aggregate functions applied to each set of products found in the account table, this query fails because you have not *explicitly* specified how the data should be grouped. Therefore, you will need to add a group by clause to specify over which group of rows the aggregate functions should be applied:

```
mysql> SELECT product_cd,
    ->    MAX(avail_balance) max_balance,
    ->    MIN(avail_balance) min_balance,
    ->    AVG(avail_balance) avg_balance,
    ->    SUM(avail_balance) tot_balance,
    ->    COUNT(*) num_accts
    -> FROM account
    -> GROUP BY product_cd;
+------------+-------------+-------------+--------------+-------------+-----------+
| product_cd | max_balance | min_balance | avg_balance  | tot_balance | num_accts |
+------------+-------------+-------------+--------------+-------------+-----------+
| BUS        |     9345.55 |        0.00 |  4672.774902 |     9345.55 |         2 |
| CD         |    10000.00 |     1500.00 |  4875.000000 |    19500.00 |         4 |
| CHK        |    38552.05 |      122.37 |  7300.800985 |    73008.01 |        10 |
| MM         |     9345.55 |     2212.50 |  5681.713216 |    17045.14 |         3 |
| SAV        |      767.77 |      200.00 |   463.940002 |     1855.76 |         4 |
| SBL        |    50000.00 |    50000.00 | 50000.000000 |    50000.00 |         1 |
+------------+-------------+-------------+--------------+-------------+-----------+
6 rows in set (0.00 sec)
```

With the inclusion of the group by clause, the server knows to group together rows having the same value in the product_cd column first and then to apply the five aggregate functions to each of the six groups.

Counting Distinct Values

When using the count() function to determine the number of members in each group, you have your choice of counting *all* members in the group, or counting only the *distinct* values for a column across all members of the group. For example, consider the following data, which shows the employee responsible for opening each account:

```
mysql> SELECT account_id, open_emp_id
    -> FROM account
    -> ORDER BY open_emp_id;
+------------+-------------+
| account_id | open_emp_id |
+------------+-------------+
|          8 |           1 |
|          9 |           1 |
|         10 |           1 |
|         12 |           1 |
|         13 |           1 |
|         17 |           1 |
|         18 |           1 |
```

```
|         19 |            1 |
|          1 |           10 |
|          2 |           10 |
|          3 |           10 |
|          4 |           10 |
|          5 |           10 |
|         14 |           10 |
|         22 |           10 |
|          6 |           13 |
|          7 |           13 |
|         24 |           13 |
|         11 |           16 |
|         15 |           16 |
|         16 |           16 |
|         20 |           16 |
|         21 |           16 |
|         23 |           16 |
+------------+--------------+
24 rows in set (0.00 sec)
```

As you can see, multiple accounts were opened by four different employees (employee IDs 1, 10, 13, and 16). Let's say that, instead of performing a manual count, you want to create a query that counts the number of employees who have opened accounts. If you apply the count() function to the open_emp_id column, you will see the following results:

```
mysql> SELECT COUNT(open_emp_id)
    -> FROM account;
+--------------------+
| COUNT(open_emp_id) |
+--------------------+
|                 24 |
+--------------------+
1 row in set (0.00 sec)
```

In this case, specifying the open_emp_id column as the column to be counted generates the same results as specifying count(*). If you want to count *distinct* values in the group rather than just counting the number of rows in the group, you need to specify the distinct keyword, as in:

```
mysql> SELECT COUNT(DISTINCT open_emp_id)
    -> FROM account;
+-----------------------------+
| COUNT(DISTINCT open_emp_id) |
+-----------------------------+
|                           4 |
+-----------------------------+
1 row in set (0.00 sec)
```

By specifying distinct, therefore, the count() function examines the values of a column for each member of the group in order to find and remove duplicates, rather than simply counting the number of values in the group.

Using Expressions

Along with using columns as arguments to aggregate functions, you can build expressions to use as arguments. For example, you may want to find the maximum value of pending deposits across all accounts, which is calculated by subtracting the available balance from the pending balance. You can achieve this via the following query:

```
mysql> SELECT MAX(pending_balance - avail_balance) max_uncleared
    -> FROM account;
+---------------+
| max_uncleared |
+---------------+
|        660.00 |
+---------------+
1 row in set (0.00 sec)
```

While this example uses a fairly simple expression, expressions used as arguments to aggregate functions can be as complex as needed, as long as they return a number, string, or date. In Chapter 11, I show you how you can use case expressions with aggregate functions to determine whether a particular row should or should not be included in an aggregation.

How Nulls Are Handled

When performing aggregations, or, indeed, any type of numeric calculation, you should always consider how null values might affect the outcome of your calculation. To illustrate, I will build a simple table to hold numeric data and populate it with the set {1, 3, 5}:

```
mysql> CREATE TABLE number_tbl
    -> (val SMALLINT);
Query OK, 0 rows affected (0.01 sec)

mysql> INSERT INTO number_tbl VALUES (1);
Query OK, 1 row affected (0.00 sec)

mysql> INSERT INTO number_tbl VALUES (3);
Query OK, 1 row affected (0.00 sec)

mysql> INSERT INTO number_tbl VALUES (5);
Query OK, 1 row affected (0.00 sec)
```

Consider the following query, which performs five aggregate functions on the set of numbers:

```
mysql> SELECT COUNT(*) num_rows,
    ->    COUNT(val) num_vals,
    ->    SUM(val) total,
    ->    MAX(val) max_val,
    ->    AVG(val) avg_val
    -> FROM number_tbl;
+----------+----------+-------+---------+---------+
```

```
| num_rows | num_vals | total | max_val | avg_val |
+----------+----------+-------+---------+---------+
|        3 |        3 |     9 |       5 |  3.0000 |
+----------+----------+-------+---------+---------+
1 row in set (0.08 sec)
```

The results are as you would expect: both count(*) and count(val) return the value 3, sum(val) returns the value 9, max(val) returns 5, and avg(val) returns 3. Next, I will add a null value to the number_tbl table and run the query again:

```
mysql> INSERT INTO number_tbl VALUES (NULL);
Query OK, 1 row affected (0.01 sec)

mysql> SELECT COUNT(*) num_rows,
    ->   COUNT(val) num_vals,
    ->   SUM(val) total,
    ->   MAX(val) max_val,
    ->   AVG(val) avg_val
    -> FROM number_tbl;
+----------+----------+-------+---------+---------+
| num_rows | num_vals | total | max_val | avg_val |
+----------+----------+-------+---------+---------+
|        4 |        3 |     9 |       5 |  3.0000 |
+----------+----------+-------+---------+---------+
1 row in set (0.00 sec)
```

Even with the addition of the null value to the table, the sum(), max(), and avg() functions all return the same values, indicating that they ignore any null values encountered. The count(*) function now returns the value 4, which is valid since the number_tbl table contains four rows, while the count(val) function still returns the value 3. The difference is that count(*) counts the number of rows, whereas count(val) counts the number of *values* contained in the val column and ignores any null values encountered.

Generating Groups

People are rarely interested in looking at raw data; instead, people engaging in data analysis will want to manipulate the raw data to better suit their needs. Examples of common data manipulations include:

- Generating totals for a geographic region, such as total European sales
- Finding outliers, such as the top salesperson for 2005
- Determining frequencies, such as the number of new accounts opened for each branch

To answer these types of queries, you will need to ask the database server to group rows together by one or more columns or expressions. As you have seen already in several examples, the group by clause is the mechanism for grouping data within a query. In this section, you will see how to group data by one or more columns, how to group data using expressions, and how to generate rollups within groups.

Single-Column Grouping

Single-column groups are the simplest and most-often-used type of grouping. If you want to find the total balances for each product, for example, you need only group on the `account.product_cd` column, as in:

```
mysql> SELECT product_cd, SUM(avail_balance) prod_balance
    -> FROM account
    -> GROUP BY product_cd;
+------------+--------------+
| product_cd | prod_balance |
+------------+--------------+
| BUS        |      9345.55 |
| CD         |     19500.00 |
| CHK        |     73008.01 |
| MM         |     17045.14 |
| SAV        |      1855.76 |
| SBL        |     50000.00 |
+------------+--------------+
6 rows in set (0.00 sec)
```

This query generates six groups, one for each product, and then sums the available balances for each member of the group.

Multicolumn Grouping

In some cases, you may want to generate groups that span *more* than one column. Expanding on the previous example, imagine that you want to find the total balances not just for each product, but for both products and branches (e.g., what's the total balance for all checking accounts opened at the Woburn branch?). The following example shows how you can accomplish this:

```
mysql> SELECT product_cd, open_branch_id,
    ->   SUM(avail_balance) tot_balance
    -> FROM account
    -> GROUP BY product_cd, open_branch_id;
+------------+----------------+-------------+
| product_cd | open_branch_id | tot_balance |
+------------+----------------+-------------+
| BUS        |              2 |     9345.55 |
| BUS        |              4 |        0.00 |
| CD         |              1 |    11500.00 |
| CD         |              2 |     8000.00 |
| CHK        |              1 |      782.16 |
| CHK        |              2 |     3315.77 |
| CHK        |              3 |     1057.75 |
| CHK        |              4 |    67852.33 |
| MM         |              1 |    14832.64 |
| MM         |              3 |     2212.50 |
| SAV        |              1 |      767.77 |
| SAV        |              2 |      700.00 |
| SAV        |              4 |      387.99 |
| SBL        |              3 |    50000.00 |
```

```
+-------------+-----------------+-------------+
```
14 rows in set (0.00 sec)

This version of the query generates 14 groups, one for each combination of product and branch found in the `account` table. Along with adding the `open_branch_id` column to the `select` clause, I also added it to the `group by` clause, since `open_branch_id` is retrieved from a table and is not generated via an aggregate function.

Grouping via Expressions

Along with using columns to group data, you can build groups based on the values generated by expressions. Consider the following query, which groups employees by the year they began working for the bank:

```
mysql> SELECT EXTRACT(YEAR FROM start_date) year,
    ->   COUNT(*) how_many
    -> FROM employee
    -> GROUP BY EXTRACT(YEAR FROM start_date);
+------+----------+
| year | how_many |
+------+----------+
| 2004 |        2 |
| 2005 |        3 |
| 2006 |        8 |
| 2007 |        3 |
| 2008 |        2 |
+------+----------+
5 rows in set (0.15 sec)
```

This query employs a fairly simple expression, which uses the `extract()` function to return only the year portion of a date, to group the rows in the `employee` table.

Generating Rollups

In "Multicolumn Grouping" on page 151, I showed an example that generates total account balances for each product and branch. Let's say, however, that along with the total balances for each product/branch combination, you also want total balances for each distinct product. You could run an additional query and merge the results, you could load the results of the query into a spreadsheet, or you could build a Perl script, Java program, or some other mechanism to take that data and perform the additional calculations. Better yet, you could use the `with rollup` option to have the database server do the work for you. Here's the revised query using `with rollup` in the `group by` clause:

```
mysql> SELECT product_cd, open_branch_id,
    ->   SUM(avail_balance) tot_balance
    -> FROM account
    -> GROUP BY product_cd, open_branch_id WITH ROLLUP;

+-------------+-----------------+-------------+
| product_cd | open_branch_id | tot_balance |
```

```
+-----------+----------------+-------------+
| BUS       |              2 |     9345.55 |
| BUS       |              4 |        0.00 |
| BUS       |           NULL |     9345.55 |
| CD        |              1 |    11500.00 |
| CD        |              2 |     8000.00 |
| CD        |           NULL |    19500.00 |
| CHK       |              1 |      782.16 |
| CHK       |              2 |     3315.77 |
| CHK       |              3 |     1057.75 |
| CHK       |              4 |    67852.33 |
| CHK       |           NULL |    73008.01 |
| MM        |              1 |    14832.64 |
| MM        |              3 |     2212.50 |
| MM        |           NULL |    17045.14 |
| SAV       |              1 |      767.77 |
| SAV       |              2 |      700.00 |
| SAV       |              4 |      387.99 |
| SAV       |           NULL |     1855.76 |
| SBL       |              3 |    50000.00 |
| SBL       |           NULL |    50000.00 |
| NULL      |           NULL |   170754.46 |
+-----------+----------------+-------------+
21 rows in set (0.02 sec)
```

There are now seven additional rows in the result set, one for each of the six distinct products and one for the grand total (all products combined). For the six product rollups, a null value is provided for the open_branch_id column, since the rollup is being performed across all branches. Looking at the third line of the output, for example, you will see that a total of $9,345.55 was deposited in BUS accounts across all branches. For the grand total row, a null value is provided for both the product_cd and open_branch_id columns; the last line of output shows a total of $170,754.46 across all products and branches.

 If you are using Oracle Database, you need to use a slightly different syntax to indicate that you want a rollup performed. The group by clause for the previous query would look as follows when using Oracle:

```
GROUP BY ROLLUP(product_cd, open_branch_id)
```

The advantage of this syntax is that it allows you to perform rollups on a subset of the columns in the group by clause. If you are grouping by columns a, b, and c, for example, you could indicate that the server should perform rollups on only b and c via the following:

```
GROUP BY a, ROLLUP(b, c)
```

If, along with totals by product, you also want to calculate totals per branch, then you can use the with cube option, which generates summary rows for *all* possible combinations of the grouping columns. Unfortunately, with cube is not available in version 6.0 of MySQL, but it is available with SQL Server and Oracle Database. Here's an example using with cube, but I have removed the mysql> prompt to show that the query cannot yet be performed with MySQL:

```
SELECT product_cd, open_branch_id,
  SUM(avail_balance) tot_balance
FROM account
GROUP BY product_cd, open_branch_id WITH CUBE;
+------------+----------------+-------------+
| product_cd | open_branch_id | tot_balance |
+------------+----------------+-------------+
| NULL       |           NULL |   170754.46 |
| NULL       |              1 |    27882.57 |
| NULL       |              2 |    21361.32 |
| NULL       |              3 |    53270.25 |
| NULL       |              4 |    68240.32 |
| BUS        |              2 |     9345.55 |
| BUS        |              4 |        0.00 |
| BUS        |           NULL |     9345.55 |
| CD         |              1 |    11500.00 |
| CD         |              2 |     8000.00 |
| CD         |           NULL |    19500.00 |
| CHK        |              1 |      782.16 |
| CHK        |              2 |     3315.77 |
| CHK        |              3 |     1057.75 |
| CHK        |              4 |    67852.33 |
| CHK        |           NULL |    73008.01 |
| MM         |              1 |    14832.64 |
| MM         |              3 |     2212.50 |
| MM         |           NULL |    17045.14 |
| SAV        |              1 |      767.77 |
| SAV        |              2 |      700.00 |
| SAV        |              4 |      387.99 |
| SAV        |           NULL |     1855.76 |
| SBL        |              3 |    50000.00 |
| SBL        |           NULL |    50000.00 |
+------------+----------------+-------------+
25 rows in set (0.02 sec)
```

Using with cube generates four more rows than the with rollup version of the query, one for each of the four branch IDs. Similar to with rollup, null values are placed in the product_cd column to indicate that a branch summary is being performed.

 Once again, if you are using Oracle Database, you need to use a slightly different syntax to indicate that you want a cube operation performed. The group by clause for the previous query would look as follows when using Oracle:

```
GROUP BY CUBE(product_cd, open_branch_id)
```

Group Filter Conditions

In Chapter 4, I introduced you to various types of filter conditions and showed how you can use them in the where clause. When grouping data, you also can apply filter conditions to the data *after* the groups have been generated. The having clause is where you should place these types of filter conditions. Consider the following example:

```
mysql> SELECT product_cd, SUM(avail_balance) prod_balance
    -> FROM account
    -> WHERE status = 'ACTIVE'
    -> GROUP BY product_cd
    -> HAVING SUM(avail_balance) >= 10000;
+------------+--------------+
| product_cd | prod_balance |
+------------+--------------+
| CD         |     19500.00 |
| CHK        |     73008.01 |
| MM         |     17045.14 |
| SBL        |     50000.00 |
+------------+--------------+
4 rows in set (0.00 sec)
```

This query has two filter conditions: one in the where clause, which filters out inactive accounts, and the other in the having clause, which filters out any product whose total available balance is less than $10,000. Thus, one of the filters acts on data *before* it is grouped, and the other filter acts on data *after* the groups have been created. If you mistakenly put both filters in the where clause, you will see the following error:

```
mysql> SELECT product_cd, SUM(avail_balance) prod_balance
    -> FROM account
    -> WHERE status = 'ACTIVE'
    ->    AND SUM(avail_balance) > 10000
    -> GROUP BY product_cd;
ERROR 1111 (HY000): Invalid use of group function
```

This query fails because you cannot include an aggregate function in a query's where clause. This is because the filters in the where clause are evaluated *before* the grouping occurs, so the server can't yet perform any functions on groups.

 When adding filters to a query that includes a group by clause, think carefully about whether the filter acts on raw data, in which case it belongs in the where clause, or on grouped data, in which case it belongs in the having clause.

You may, however, include aggregate functions in the having clause, that do *not* appear in the select clause, as demonstrated by the following:

```
mysql> SELECT product_cd, SUM(avail_balance) prod_balance
    -> FROM account
    -> WHERE status = 'ACTIVE'
    -> GROUP BY product_cd
    -> HAVING MIN(avail_balance) >= 1000
```

```
    ->   AND MAX(avail_balance) <= 10000;
+------------+--------------+
| product_cd | prod_balance |
+------------+--------------+
| CD         |     19500.00 |
| MM         |     17045.14 |
+------------+--------------+
2 rows in set (0.00 sec)
```

This query generates total balances for each active product, but then the filter condition in the having clause excludes all products for which the minimum balance is less than $1,000 or the maximum balance is greater than $10,000.

Test Your Knowledge

Work through the following exercises to test your grasp of SQL's grouping and aggregating features. Check your work with the answers in Appendix C.

Exercise 8-1

Construct a query that counts the number of rows in the account table.

Exercise 8-2

Modify your query from Exercise 8-1 to count the number of accounts held by each customer. Show the customer ID and the number of accounts for each customer.

Exercise 8-3

Modify your query from Exercise 8-2 to include only those customers having at least two accounts.

Exercise 8-4 (Extra Credit)

Find the total available balance by product and branch where there is more than one account per product and branch. Order the results by total balance (highest to lowest).

CHAPTER 9

Subqueries

Subqueries are a powerful tool that you can use in all four SQL data statements. This chapter explores in great detail the many uses of the subquery.

What Is a Subquery?

A *subquery* is a query contained within another SQL statement (which I refer to as the *containing statement* for the rest of this discussion). A subquery is always enclosed within parentheses, and it is usually executed prior to the containing statement. Like any query, a subquery returns a result set that may consist of:

- A single row with a single column
- Multiple rows with a single column
- Multiple rows and columns

The type of result set the subquery returns determines how it may be used and which operators the containing statement may use to interact with the data the subquery returns. When the containing statement has finished executing, the data returned by any subqueries is discarded, making a subquery act like a temporary table with *statement scope* (meaning that the server frees up any memory allocated to the subquery results after the SQL statement has finished execution).

You already saw several examples of subqueries in earlier chapters, but here's a simple example to get started:

```
mysql> SELECT account_id, product_cd, cust_id, avail_balance
    -> FROM account
    -> WHERE account_id = (SELECT MAX(account_id) FROM account);
+------------+------------+---------+---------------+
| account_id | product_cd | cust_id | avail_balance |
+------------+------------+---------+---------------+
|         29 | SBL        |      13 |      50000.00 |
+------------+------------+---------+---------------+
1 row in set (0.65 sec)
```

In this example, the subquery returns the maximum value found in the `account_id` column in the `account` table, and the containing statement then returns data about that account. If you are ever confused about what a subquery is doing, you can run the subquery by itself (without the parentheses) to see what it returns. Here's the subquery from the previous example:

```
mysql> SELECT MAX(account_id) FROM account;
+-----------------+
| MAX(account_id) |
+-----------------+
|              29 |
+-----------------+
1 row in set (0.00 sec)
```

So, the subquery returns a single row with a single column, which allows it to be used as one of the expressions in an equality condition (if the subquery returned two or more rows, it could be *compared* to something but could not be *equal* to anything, but more on this later). In this case, you can take the value the subquery returned and substitute it into the righthand expression of the filter condition in the containing query, as in:

```
mysql> SELECT account_id, product_cd, cust_id, avail_balance
    -> FROM account
    -> WHERE account_id = 29;
+------------+------------+---------+---------------+
| account_id | product_cd | cust_id | avail_balance |
+------------+------------+---------+---------------+
|         29 | SBL        |      13 |      50000.00 |
+------------+------------+---------+---------------+
1 row in set (0.02 sec)
```

The subquery is useful in this case because it allows you to retrieve information about the highest numbered account in a single query, rather than retrieving the maximum `account_id` using one query and then writing a second query to retrieve the desired data from the `account` table. As you will see, subqueries are useful in many other situations as well, and may become one of the most powerful tools in your SQL toolkit.

Subquery Types

Along with the differences noted previously regarding the type of result set a subquery returns (single row/column, single row/multicolumn, or multiple columns), you can use another factor to differentiate subqueries; some subqueries are completely self-contained (called *noncorrelated subqueries*), while others reference columns from the containing statement (called *correlated subqueries*). The next several sections explore these two subquery types and show the different operators that you can employ to interact with them.

Noncorrelated Subqueries

The example from earlier in the chapter is a noncorrelated subquery; it may be executed alone and does not reference anything from the containing statement. Most subqueries that you encounter will be of this type unless you are writing update or delete statements, which frequently make use of correlated subqueries (more on this later). Along with being noncorrelated, the example from earlier in the chapter also returns a table comprising a single row and column. This type of subquery is known as a *scalar subquery* and can appear on either side of a condition using the usual operators (=, <>, <, >, <=, >=). The next example shows how you can use a scalar subquery in an inequality condition:

```
mysql> SELECT account_id, product_cd, cust_id, avail_balance
    -> FROM account
    -> WHERE open_emp_id <> (SELECT e.emp_id
    ->    FROM employee e INNER JOIN branch b
    ->      ON e.assigned_branch_id = b.branch_id
    ->    WHERE e.title = 'Head Teller' AND b.city = 'Woburn');
+------------+------------+---------+---------------+
| account_id | product_cd | cust_id | avail_balance |
+------------+------------+---------+---------------+
|          7 | CHK        |       3 |       1057.75 |
|          8 | MM         |       3 |       2212.50 |
|         10 | CHK        |       4 |        534.12 |
|         11 | SAV        |       4 |        767.77 |
|         12 | MM         |       4 |       5487.09 |
|         13 | CHK        |       5 |       2237.97 |
|         14 | CHK        |       6 |        122.37 |
|         15 | CD         |       6 |      10000.00 |
|         18 | CHK        |       8 |       3487.19 |
|         19 | SAV        |       8 |        387.99 |
|         21 | CHK        |       9 |        125.67 |
|         22 | MM         |       9 |       9345.55 |
|         23 | CD         |       9 |       1500.00 |
|         24 | CHK        |      10 |      23575.12 |
|         25 | BUS        |      10 |          0.00 |
|         28 | CHK        |      12 |      38552.05 |
|         29 | SBL        |      13 |      50000.00 |
+------------+------------+---------+---------------+
17 rows in set (0.86 sec)
```

This query returns data concerning all accounts that were *not* opened by the head teller at the Woburn branch (the subquery is written using the assumption that there is only a single head teller at each branch). The subquery in this example is a bit more complex than in the previous example, in that it joins two tables and includes two filter conditions. Subqueries may be as simple or as complex as you need them to be, and they may utilize any and all the available query clauses (select, from, where, group by, having, and order by).

If you use a subquery in an equality condition, but the subquery returns more than one row, you will receive an error. For example, if you modify the previous query such that the subquery returns *all* tellers at the Woburn branch instead of the single head teller, you will receive the following error:

```
mysql> SELECT account_id, product_cd, cust_id, avail_balance
    -> FROM account
    -> WHERE open_emp_id <> (SELECT e.emp_id
    ->   FROM employee e INNER JOIN branch b
    ->     ON e.assigned_branch_id = b.branch_id
    ->   WHERE e.title = 'Teller' AND b.city = 'Woburn');
ERROR 1242 (21000): Subquery returns more than 1 row
```

If you run the subquery by itself, you will see the following results:

```
mysql> SELECT e.emp_id
    -> FROM employee e INNER JOIN branch b
    ->   ON e.assigned_branch_id = b.branch_id
    -> WHERE e.title = 'Teller' AND b.city = 'Woburn';
+--------+
| emp_id |
+--------+
|     11 |
|     12 |
+--------+
2 rows in set (0.02 sec)
```

The containing query fails because an expression (open_emp_id) cannot be equated to a set of expressions (emp_ids 11 and 12). In other words, a single thing cannot be equated to a set of things. In the next section, you will see how to fix the problem by using a different operator.

Multiple-Row, Single-Column Subqueries

If your subquery returns more than one row, you will not be able to use it on one side of an equality condition, as the previous example demonstrated. However, there are four additional operators that you can use to build conditions with these types of subqueries.

The in and not in operators

While you can't *equate* a single value to a set of values, you can check to see whether a single value can be found *within* a set of values. The next example, while it doesn't use a subquery, demonstrates how to build a condition that uses the in operator to search for a value within a set of values:

```
mysql> SELECT branch_id, name, city
    -> FROM branch
    -> WHERE name IN ('Headquarters', 'Quincy Branch');
+-----------+---------------+---------+
| branch_id | name          | city    |
+-----------+---------------+---------+
|         1 | Headquarters  | Waltham |
|         3 | Quincy Branch | Quincy  |
```

```
+-----------+---------------+---------+
```
2 rows in set (0.03 sec)

The expression on the lefthand side of the condition is the name column, while the righthand side of the condition is a set of strings. The in operator checks to see whether either of the strings can be found in the name column; if so, the condition is met and the row is added to the result set. You could achieve the same results using two equality conditions, as in:

```
mysql> SELECT branch_id, name, city
    -> FROM branch
    -> WHERE name = 'Headquarters' OR name = 'Quincy Branch';
+-----------+---------------+---------+
| branch_id | name          | city    |
+-----------+---------------+---------+
|         1 | Headquarters  | Waltham |
|         3 | Quincy Branch | Quincy  |
+-----------+---------------+---------+
2 rows in set (0.01 sec)
```

While this approach seems reasonable when the set contains only two expressions, it is easy to see why a single condition using the in operator would be preferable if the set contained dozens (or hundreds, thousands, etc.) of values.

Although you will occasionally create a set of strings, dates, or numbers to use on one side of a condition, you are more likely to generate the set at query execution via a subquery that returns one or more rows. The following query uses the in operator with a subquery on the righthand side of the filter condition to see which employees supervise other employees:

```
mysql> SELECT emp_id, fname, lname, title
    -> FROM employee
    -> WHERE emp_id IN (SELECT superior_emp_id
    ->    FROM employee);
+--------+---------+-----------+--------------------+
| emp_id | fname   | lname     | title              |
+--------+---------+-----------+--------------------+
|      1 | Michael | Smith     | President          |
|      3 | Robert  | Tyler     | Treasurer          |
|      4 | Susan   | Hawthorne | Operations Manager |
|      6 | Helen   | Fleming   | Head Teller        |
|     10 | Paula   | Roberts   | Head Teller        |
|     13 | John    | Blake     | Head Teller        |
|     16 | Theresa | Markham   | Head Teller        |
+--------+---------+-----------+--------------------+
7 rows in set (0.01 sec)
```

The subquery returns the IDs of all employees who supervise other employees, and the containing query retrieves four columns from the employee table for these employees. Here are the results of the subquery:

```
mysql> SELECT superior_emp_id
    -> FROM employee;
+-----------------+
```

```
| superior_emp_id |
+-----------------+
|            NULL |
|               1 |
|               1 |
|               3 |
|               4 |
|               4 |
|               4 |
|               4 |
|               6 |
|               6 |
|               6 |
|              10 |
|              10 |
|              13 |
|              13 |
|              16 |
|              16 |
+-----------------+
18 rows in set (0.00 sec)
```

As you can see, some employee IDs are listed more than once, since some employees supervise multiple people. This doesn't adversely affect the results of the containing query, since it doesn't matter whether an employee ID can be found in the result set of the subquery once or more than once. Of course, you could add the **distinct** keyword to the subquery's **select** clause if it bothers you to have duplicates in the table returned by the subquery, but it won't change the containing query's result set.

Along with seeing whether a value exists within a set of values, you can check the converse using the not in operator. Here's another version of the previous query using not in instead of in:

```
mysql> SELECT emp_id, fname, lname, title
    -> FROM employee
    -> WHERE emp_id NOT IN (SELECT superior_emp_id
    ->   FROM employee
    ->   WHERE superior_emp_id IS NOT NULL);
+--------+----------+----------+----------------+
| emp_id | fname    | lname    | title          |
+--------+----------+----------+----------------+
|      2 | Susan    | Barker   | Vice President |
|      5 | John     | Gooding  | Loan Manager   |
|      7 | Chris    | Tucker   | Teller         |
|      8 | Sarah    | Parker   | Teller         |
|      9 | Jane     | Grossman | Teller         |
|     11 | Thomas   | Ziegler  | Teller         |
|     12 | Samantha | Jameson  | Teller         |
|     14 | Cindy    | Mason    | Teller         |
|     15 | Frank    | Portman  | Teller         |
|     17 | Beth     | Fowler   | Teller         |
|     18 | Rick     | Tulman   | Teller         |
```

```
+--------+----------+----------+----------------+
11 rows in set (0.00 sec)
```

This query finds all employees who do *not* supervise other people. For this query, I needed to add a filter condition to the subquery to ensure that null values do not appear in the table returned by the subquery; see the next section for an explanation of why this filter is needed in this case.

The all operator

While the in operator is used to see whether an expression can be found within a set of expressions, the all operator allows you to make comparisons between a single value and every value in a set. To build such a condition, you will need to use one of the comparison operators (=, <>, <, >, etc.) in conjunction with the all operator. For example, the next query finds all employees whose employee IDs are not equal to any of the supervisor employee IDs:

```
mysql> SELECT emp_id, fname, lname, title
    -> FROM employee
    -> WHERE emp_id <> ALL (SELECT superior_emp_id
    ->    FROM employee
    ->    WHERE superior_emp_id IS NOT NULL);
+--------+----------+----------+----------------+
| emp_id | fname    | lname    | title          |
+--------+----------+----------+----------------+
|      2 | Susan    | Barker   | Vice President |
|      5 | John     | Gooding  | Loan Manager   |
|      7 | Chris    | Tucker   | Teller         |
|      8 | Sarah    | Parker   | Teller         |
|      9 | Jane     | Grossman | Teller         |
|     11 | Thomas   | Ziegler  | Teller         |
|     12 | Samantha | Jameson  | Teller         |
|     14 | Cindy    | Mason    | Teller         |
|     15 | Frank    | Portman  | Teller         |
|     17 | Beth     | Fowler   | Teller         |
|     18 | Rick     | Tulman   | Teller         |
+--------+----------+----------+----------------+
11 rows in set (0.05 sec)
```

Once again, the subquery returns the set of IDs for those employees who supervise other people, and the containing query returns data for each employee whose ID is not equal to all of the IDs returned by the subquery. In other words, the query finds all employees who are not supervisors. If this approach seems a bit clumsy to you, you are in good company; most people would prefer to phrase the query differently and avoid using the all operator. For example, this query generates the same results as the last example in the previous section, which used the not in operator. It's a matter of preference, but I think that most people would find the version that uses not in to be easier to understand.

When using not in or <> all to compare a value to a set of values, you must be careful to ensure that the set of values does not contain a null value, because the server equates the value on the lefthand side of the expression to each member of the set, and any attempt to equate a value to null yields unknown. Thus, the following query returns an empty set:

```
mysql> SELECT emp_id, fname, lname, title
    -> FROM employee
    -> WHERE emp_id NOT IN (1, 2, NULL);
Empty set (0.00 sec)
```

In some cases, the all operator is a bit more natural. The next example uses all to find accounts having an available balance smaller than all of Frank Tucker's accounts:

```
mysql> SELECT account_id, cust_id, product_cd, avail_balance
    -> FROM account
    -> WHERE avail_balance < ALL (SELECT a.avail_balance
    ->   FROM account a INNER JOIN individual i
    ->     ON a.cust_id = i.cust_id
    ->   WHERE i.fname = 'Frank' AND i.lname = 'Tucker');
```

account_id	cust_id	product_cd	avail_balance
2	1	SAV	500.00
5	2	SAV	200.00
10	4	CHK	534.12
11	4	SAV	767.77
14	6	CHK	122.37
19	8	SAV	387.99
21	9	CHK	125.67
25	10	BUS	0.00

```
8 rows in set (0.17 sec)
```

Here's the data returned by the subquery, which consists of the available balance from each of Frank's accounts:

```
mysql> SELECT a.avail_balance
    -> FROM account a INNER JOIN individual i
    ->   ON a.cust_id = i.cust_id
    -> WHERE i.fname = 'Frank' AND i.lname = 'Tucker';
```

avail_balance
1057.75
2212.50

```
2 rows in set (0.01 sec)
```

Frank has two accounts, with the lowest balance being $1,057.75. The containing query finds all accounts having a balance smaller than any of Frank's accounts, so the result set includes all accounts having a balance less than $1,057.75.

The any operator

Like the all operator, the any operator allows a value to be compared to the members of a set of values; unlike all, however, a condition using the any operator evaluates to true as soon as a single comparison is favorable. This is different from the previous example using the all operator, which evaluates to true only if comparisons against *all* members of the set are favorable. For example, you might want to find all accounts having an available balance greater than *any* of Frank Tucker's accounts:

```
mysql> SELECT account_id, cust_id, product_cd, avail_balance
    -> FROM account
    -> WHERE avail_balance > ANY (SELECT a.avail_balance
    ->   FROM account a INNER JOIN individual i
    ->     ON a.cust_id = i.cust_id
    ->   WHERE i.fname = 'Frank' AND i.lname = 'Tucker');
+------------+---------+------------+---------------+
| account_id | cust_id | product_cd | avail_balance |
+------------+---------+------------+---------------+
|          3 |       1 | CD         |       3000.00 |
|          4 |       2 | CHK        |       2258.02 |
|          8 |       3 | MM         |       2212.50 |
|         12 |       4 | MM         |       5487.09 |
|         13 |       5 | CHK        |       2237.97 |
|         15 |       6 | CD         |      10000.00 |
|         17 |       7 | CD         |       5000.00 |
|         18 |       8 | CHK        |       3487.19 |
|         22 |       9 | MM         |       9345.55 |
|         23 |       9 | CD         |       1500.00 |
|         24 |      10 | CHK        |      23575.12 |
|         27 |      11 | BUS        |       9345.55 |
|         28 |      12 | CHK        |      38552.05 |
|         29 |      13 | SBL        |      50000.00 |
+------------+---------+------------+---------------+
14 rows in set (0.00 sec)
```

Frank has two accounts with balances of $1,057.75 and $2,212.50; to have a balance greater than *any* of these two accounts, an account must have a balance of at least $1,057.75.

> Although most people prefer to use in, using = any is equivalent to using the in operator.

Multicolumn Subqueries

So far, all of the subquery examples in this chapter have returned a single column and one or more rows. In certain situations, however, you can use subqueries that return two or more columns. To show the utility of multiple-column subqueries, it might help to look first at an example that uses multiple, single-column subqueries:

```
mysql> SELECT account_id, product_cd, cust_id
    -> FROM account
    -> WHERE open_branch_id = (SELECT branch_id
    ->    FROM branch
    ->    WHERE name = 'Woburn Branch')
    ->    AND open_emp_id IN (SELECT emp_id
    ->    FROM employee
    ->    WHERE title = 'Teller' OR title = 'Head Teller');
+------------+------------+---------+
| account_id | product_cd | cust_id |
+------------+------------+---------+
|          1 | CHK        |       1 |
|          2 | SAV        |       1 |
|          3 | CD         |       1 |
|          4 | CHK        |       2 |
|          5 | SAV        |       2 |
|         17 | CD         |       7 |
|         27 | BUS        |      11 |
+------------+------------+---------+
7 rows in set (0.09 sec)
```

This query uses two subqueries to identify the ID of the Woburn branch and the IDs of all bank tellers, and the containing query then uses this information to retrieve all checking accounts opened by a teller at the Woburn branch. However, since the employee table includes information about which branch each employee is assigned to, you can achieve the same results by comparing both the account.open_branch_id and account.open_emp_id columns to a single subquery against the employee and branch tables. To do so, your filter condition must name both columns from the account table surrounded by parentheses and in the same order as returned by the subquery, as in:

```
mysql> SELECT account_id, product_cd, cust_id
    -> FROM account
    -> WHERE (open_branch_id, open_emp_id) IN
    ->   (SELECT b.branch_id, e.emp_id
    ->    FROM branch b INNER JOIN employee e
    ->      ON b.branch_id = e.assigned_branch_id
    ->    WHERE b.name = 'Woburn Branch'
    ->      AND (e.title = 'Teller' OR e.title = 'Head Teller'));
+------------+------------+---------+
| account_id | product_cd | cust_id |
+------------+------------+---------+
|          1 | CHK        |       1 |
|          2 | SAV        |       1 |
|          3 | CD         |       1 |
|          4 | CHK        |       2 |
|          5 | SAV        |       2 |
|         17 | CD         |       7 |
|         27 | BUS        |      11 |
+------------+------------+---------+
7 rows in set (0.00 sec)
```

This version of the query performs the same function as the previous example, but with a single subquery that returns two columns instead of two subqueries that each return a single column.

Of course, you could rewrite the previous example simply to join the three tables instead of using a subquery, but it's helpful when learning SQL to see multiple ways of achieving the same results. Here's another example, however, that requires a subquery. Let's say that there have been some customer complaints regarding incorrect values in the available/pending balance columns in the `account` table. Your job is to find all accounts whose balances don't match the sum of the transaction amounts for that account. Here's a partial solution to the problem:

```
SELECT 'ALERT! : Account #1 Has Incorrect Balance!'
FROM account
WHERE (avail_balance, pending_balance) <>
 (SELECT SUM(<expression to generate available balance>),
    SUM(<expression to generate pending balance>)
  FROM transaction
  WHERE account_id = 1)
  AND account_id = 1;
```

As you can see, I have neglected to fill in the expressions used to sum the transaction amounts for the available and pending balance calculations, but I promise to finish the job in Chapter 11 after you learn how to build `case` expressions. Even so, the query is complete enough to see that the subquery is generating two sums from the `transaction` table that are then compared to the `avail_balance` and `pending_balance` columns in the `account` table. Both the subquery and the containing query include the filter condition `account_id = 1`, so the query in its present form will check only a single account at a time. In the next section, you will learn how to write a more general form of the query that will check *all* accounts with a single execution.

Correlated Subqueries

All of the subqueries shown thus far have been independent of their containing statements, meaning that you can execute them by themselves and inspect the results. A *correlated subquery*, on the other hand, is *dependent* on its containing statement from which it references one or more columns. Unlike a noncorrelated subquery, a correlated subquery is not executed once prior to execution of the containing statement; instead, the correlated subquery is executed once for each candidate row (rows that might be included in the final results). For example, the following query uses a correlated subquery to count the number of accounts for each customer, and the containing query then retrieves those customers having exactly two accounts:

```
mysql> SELECT c.cust_id, c.cust_type_cd, c.city
    -> FROM customer c
    -> WHERE 2 = (SELECT COUNT(*)
    ->    FROM account a
    ->    WHERE a.cust_id = c.cust_id);
+---------+--------------+---------+
| cust_id | cust_type_cd | city    |
+---------+--------------+---------+
|       2 | I            | Woburn  |
|       3 | I            | Quincy  |
```

```
|       6 | I          |  Waltham |
|       8 | I          |  Salem   |
|      10 | B          |  Salem   |
+---------+------------+----------+
5 rows in set (0.01 sec)
```

The reference to `c.cust_id` at the very end of the subquery is what makes the subquery correlated; the containing query must supply values for `c.cust_id` for the subquery to execute. In this case, the containing query retrieves all 13 rows from the `customer` table and executes the subquery once for each customer, passing in the appropriate customer ID for each execution. If the subquery returns the value 2, then the filter condition is met and the row is added to the result set.

Along with equality conditions, you can use correlated subqueries in other types of conditions, such as the range condition illustrated here:

```
mysql> SELECT c.cust_id, c.cust_type_cd, c.city
    -> FROM customer c
    -> WHERE (SELECT SUM(a.avail_balance)
    ->        FROM account a
    ->        WHERE a.cust_id = c.cust_id)
    ->    BETWEEN 5000 AND 10000;
+---------+--------------+------------+
| cust_id | cust_type_cd | city       |
+---------+--------------+------------+
|       4 | I            | Waltham    |
|       7 | I            | Wilmington |
|      11 | B            | Wilmington |
+---------+--------------+------------+
3 rows in set (0.02 sec)
```

This variation on the previous query finds all customers whose total available balance across all accounts lies between $5,000 and $10,000. Once again, the correlated subquery is executed 13 times (once for each customer row), and each execution of the subquery returns the total account balance for the given customer.

 Another subtle difference in the previous query is that the subquery is on the lefthand side of the condition, which may look a bit odd but is perfectly valid.

At the end of the previous section, I demonstrated how to check the available and pending balances of an account against the transactions logged against the account, and I promised to show you how to modify the example to run all accounts in a single execution. Here's the example again:

```
SELECT 'ALERT! : Account #1 Has Incorrect Balance!'
FROM account
WHERE (avail_balance, pending_balance) <>
  (SELECT SUM(<expression to generate available balance>),
     SUM(<expression to generate pending balance>)
   FROM transaction
```

```
    WHERE account_id = 1)
    AND account_id = 1;
```

Using a correlated subquery instead of a noncorrelated subquery, you can execute the containing query once, and the subquery will be run for each account. Here's the updated version:

```
SELECT CONCAT('ALERT! : Account #', a.account_id,
  ' Has Incorrect Balance!')
FROM account a
WHERE (a.avail_balance, a.pending_balance) <>
  (SELECT SUM(<expression to generate available balance>),
    SUM(<expression to generate pending balance>)
  FROM transaction t
  WHERE t.account_id = a.account_id);
```

The subquery now includes a filter condition linking the transaction's account ID to the account ID from the containing query. The select clause has also been modified to concatenate an alert message that includes the account ID rather than the hardcoded value 1.

The exists Operator

While you will often see correlated subqueries used in equality and range conditions, the most common operator used to build conditions that utilize correlated subqueries is the exists operator. You use the exists operator when you want to identify that a relationship exists without regard for the quantity; for example, the following query finds all the accounts for which a transaction was posted on a particular day, without regard for how many transactions were posted:

```
SELECT a.account_id, a.product_cd, a.cust_id, a.avail_balance
FROM account a
WHERE EXISTS (SELECT 1
  FROM transaction t
  WHERE t.account_id = a.account_id
    AND t.txn_date = '2008-09-22');
```

Using the exists operator, your subquery can return zero, one, or many rows, and the condition simply checks whether the subquery returned any rows. If you look at the select clause of the subquery, you will see that it consists of a single literal (1); since the condition in the containing query only needs to know how many rows have been returned, the actual data the subquery returned is irrelevant. Your subquery can return whatever strikes your fancy, as demonstrated next:

```
SELECT a.account_id, a.product_cd, a.cust_id, a.avail_balance
FROM account a
WHERE EXISTS (SELECT t.txn_id, 'hello', 3.1415927
  FROM transaction t
  WHERE t.account_id = a.account_id
    AND t.txn_date = '2008-09-22');
```

However, the convention is to specify either select 1 or select * when using exists.

You may also use `not exists` to check for subqueries that return no rows, as demonstrated by the following:

```
mysql> SELECT a.account_id, a.product_cd, a.cust_id
    -> FROM account a
    -> WHERE NOT EXISTS (SELECT 1
    ->    FROM business b
    ->    WHERE b.cust_id = a.cust_id);
+------------+------------+---------+
| account_id | product_cd | cust_id |
+------------+------------+---------+
|          1 | CHK        |       1 |
|          2 | SAV        |       1 |
|          3 | CD         |       1 |
|          4 | CHK        |       2 |
|          5 | SAV        |       2 |
|          7 | CHK        |       3 |
|          8 | MM         |       3 |
|         10 | CHK        |       4 |
|         11 | SAV        |       4 |
|         12 | MM         |       4 |
|         13 | CHK        |       5 |
|         14 | CHK        |       6 |
|         15 | CD         |       6 |
|         17 | CD         |       7 |
|         18 | CHK        |       8 |
|         19 | SAV        |       8 |
|         21 | CHK        |       9 |
|         22 | MM         |       9 |
|         23 | CD         |       9 |
+------------+------------+---------+
19 rows in set (0.99 sec)
```

This query finds all customers whose customer ID does not appear in the `business` table, which is a roundabout way of finding all nonbusiness customers.

Data Manipulation Using Correlated Subqueries

All of the examples thus far in the chapter have been `select` statements, but don't think that means that subqueries aren't useful in other SQL statements. Subqueries are used heavily in `update`, `delete`, and `insert` statements as well, with correlated subqueries appearing frequently in `update` and `delete` statements. Here's an example of a correlated subquery used to modify the `last_activity_date` column in the `account` table:

```
UPDATE account a
SET a.last_activity_date =
 (SELECT MAX(t.txn_date)
  FROM transaction t
  WHERE t.account_id = a.account_id);
```

This statement modifies every row in the `account` table (since there is no `where` clause) by finding the latest transaction date for each account. While it seems reasonable to expect that every account will have at least one transaction linked to it, it would be best to check whether an account has any transactions before attempting to update the last_activity_date column; otherwise, the column will be set to null, since the subquery would return no rows. Here's another version of the update statement, this time employing a where clause with a second correlated subquery:

```
UPDATE account a
SET a.last_activity_date =
 (SELECT MAX(t.txn_date)
  FROM transaction t
  WHERE t.account_id = a.account_id)
WHERE EXISTS (SELECT 1
  FROM transaction t
  WHERE t.account_id = a.account_id);
```

The two correlated subqueries are identical except for the select clauses. The subquery in the set clause, however, executes only if the condition in the update statement's where clause evaluates to true (meaning that at least one transaction was found for the account), thus protecting the data in the last_activity_date column from being overwritten with a null.

Correlated subqueries are also common in delete statements. For example, you may run a data maintenance script at the end of each month that removes unnecessary data. The script might include the following statement, which removes data from the department table that has no child rows in the employee table:

```
DELETE FROM department
WHERE NOT EXISTS (SELECT 1
  FROM employee
  WHERE employee.dept_id = department.dept_id);
```

When using correlated subqueries with delete statements in MySQL, keep in mind that, for whatever reason, table aliases are not allowed when using delete, which is why I had to use the entire table name in the subquery. With most other database servers, you could provide aliases for the department and employee tables, such as:

```
DELETE FROM department d
WHERE NOT EXISTS (SELECT 1
  FROM employee e
  WHERE e.dept_id = d.dept_id);
```

When to Use Subqueries

Now that you have learned about the different types of subqueries and the different operators that you can employ to interact with the data returned by subqueries, it's time to explore the many ways in which you can use subqueries to build powerful SQL statements. The next three sections demonstrate how you may use subqueries to construct custom tables, to build conditions, and to generate column values in result sets.

Subqueries As Data Sources

Back in Chapter 3, I stated that the from clause of a select statement names the *tables* to be used by the query. Since a subquery generates a result set containing rows and columns of data, it is perfectly valid to include subqueries in your from clause along with tables. Although it might, at first glance, seem like an interesting feature without much practical merit, using subqueries alongside tables is one of the most powerful tools available when writing queries. Here's a simple example:

```
mysql> SELECT d.dept_id, d.name, e_cnt.how_many num_employees
    -> FROM department d INNER JOIN
    ->  (SELECT dept_id, COUNT(*) how_many
    ->   FROM employee
    ->   GROUP BY dept_id) e_cnt
    ->   ON d.dept_id = e_cnt.dept_id;
+---------+----------------+---------------+
| dept_id | name           | num_employees |
+---------+----------------+---------------+
|       1 | Operations     |            14 |
|       2 | Loans          |             1 |
|       3 | Administration |             3 |
+---------+----------------+---------------+
3 rows in set (0.04 sec)
```

In this example, a subquery generates a list of department IDs along with the number of employees assigned to each department. Here's the result set generated by the subquery:

```
mysql> SELECT dept_id, COUNT(*) how_many
    -> FROM employee
    -> GROUP BY dept_id;
+---------+----------+
| dept_id | how_many |
+---------+----------+
|       1 |       14 |
|       2 |        1 |
|       3 |        3 |
+---------+----------+
3 rows in set (0.00 sec)
```

The subquery is given the name e_cnt and is joined to the department table via the dept_id column. The containing query then retrieves the department ID and name from the department table, along with the employee count from the e_cnt subquery.

Subqueries used in the from clause must be noncorrelated; they are executed first, and the data is held in memory until the containing query finishes execution. Subqueries offer immense flexibility when writing queries, because you can go far beyond the set of available tables to create virtually any view of the data that you desire, and then join the results to other tables or subqueries. If you are writing reports or generating data feeds to external systems, you may be able to do things with a single query that used to demand multiple queries or a procedural language to accomplish.

Data fabrication

Along with using subqueries to summarize existing data, you can use subqueries to generate data that doesn't exist in any form within your database. For example, you may wish to group your customers by the amount of money held in deposit accounts, but you want to use group definitions that are not stored in your database. For example, let's say you want to sort your customers into the groups shown in Table 9-1.

Table 9-1. Customer balance groups

Group name	Lower limit	Upper limit
Small Fry	0	$4,999.99
Average Joes	$5,000	$9,999.99
Heavy Hitters	$10,000	$9,999,999.99

To generate these groups within a single query, you will need a way to define these three groups. The first step is to define a query that generates the group definitions:

```
mysql> SELECT 'Small Fry' name, 0 low_limit, 4999.99 high_limit
    -> UNION ALL
    -> SELECT 'Average Joes' name, 5000 low_limit, 9999.99 high_limit
    -> UNION ALL
    -> SELECT 'Heavy Hitters' name, 10000 low_limit, 9999999.99 high_limit;
+---------------+-----------+-------------+
| name          | low_limit | high_limit  |
+---------------+-----------+-------------+
| Small Fry     |         0 |     4999.99 |
| Average Joes  |      5000 |     9999.99 |
| Heavy Hitters |     10000 | 9999999.99  |
+---------------+-----------+-------------+
3 rows in set (0.00 sec)
```

I have used the set operator union all to merge the results from three separate queries into a single result set. Each query retrieves three literals, and the results from the three queries are put together to generate a result set with three rows and three columns. You now have a query to generate the desired groups, and you can place it into the from clause of another query to generate your customer groups:

```
mysql> SELECT groups.name, COUNT(*) num_customers
    -> FROM
    ->  (SELECT SUM(a.avail_balance) cust_balance
    ->   FROM account a INNER JOIN product p
    ->     ON a.product_cd = p.product_cd
    ->   WHERE p.product_type_cd = 'ACCOUNT'
    ->   GROUP BY a.cust_id) cust_rollup
    ->   INNER JOIN
    ->  (SELECT 'Small Fry' name, 0 low_limit, 4999.99 high_limit
    ->   UNION ALL
    ->   SELECT 'Average Joes' name, 5000 low_limit,
    ->     9999.99 high_limit
    ->   UNION ALL
    ->   SELECT 'Heavy Hitters' name, 10000 low_limit,
```

```
->        9999999.99 high_limit) groups
->     ON cust_rollup.cust_balance
->        BETWEEN groups.low_limit AND groups.high_limit
-> GROUP BY groups.name;
+----------------+----------------+
| name           | num_customers  |
+----------------+----------------+
| Average Joes   |             2  |
| Heavy Hitters  |             4  |
| Small Fry      |             5  |
+----------------+----------------+
3 rows in set (0.01 sec)
```

The from clause contains two subqueries; the first subquery, named cust_rollup, returns the total deposit balances for each customer, while the second subquery, named groups, generates the three customer groupings. Here's the data generated by cust_rollup:

```
mysql> SELECT SUM(a.avail_balance) cust_balance
    -> FROM account a INNER JOIN product p
    ->    ON a.product_cd = p.product_cd
    -> WHERE p.product_type_cd = 'ACCOUNT'
    -> GROUP BY a.cust_id;
+--------------+
| cust_balance |
+--------------+
|      4557.75 |
|      2458.02 |
|      3270.25 |
|      6788.98 |
|      2237.97 |
|     10122.37 |
|      5000.00 |
|      3875.18 |
|     10971.22 |
|     23575.12 |
|     38552.05 |
+--------------+
11 rows in set (0.05 sec)
```

The data generated by cust_rollup is then joined to the groups table via a range condition (cust_rollup.cust_balance BETWEEN groups.low_limit AND groups.high_limit). Finally, the joined data is grouped and the number of customers in each group is counted to generate the final result set.

Of course, you could simply decide to build a permanent table to hold the group definitions instead of using a subquery. Using that approach, you would find your database to be littered with small special-purpose tables after awhile, and you wouldn't remember the reason for which most of them were created. I've worked in environments where the database users were allowed to create their own tables for special purposes, and the results were disastrous (tables not included in backups, tables lost during server upgrades, server downtime due to space allocation issues, etc.). Armed with subqueries, however, you will be able to adhere to a policy where tables are added to a database only when there is a clear business need to store new data.

Task-oriented subqueries

In systems used for reporting or data-feed generation, you will often come across queries such as the following:

```
mysql> SELECT p.name product, b.name branch,
    ->   CONCAT(e.fname, ' ', e.lname) name,
    ->   SUM(a.avail_balance) tot_deposits
    -> FROM account a INNER JOIN employee e
    ->   ON a.open_emp_id = e.emp_id
    ->   INNER JOIN branch b
    ->   ON a.open_branch_id = b.branch_id
    ->   INNER JOIN product p
    ->   ON a.product_cd = p.product_cd
    -> WHERE p.product_type_cd = 'ACCOUNT'
    -> GROUP BY p.name, b.name, e.fname, e.lname
    -> ORDER BY 1,2;
+------------------------+----------------+-----------------+--------------+
| product                | branch         | name            | tot_deposits |
+------------------------+----------------+-----------------+--------------+
| certificate of deposit | Headquarters   | Michael Smith   |     11500.00 |
| certificate of deposit | Woburn Branch  | Paula Roberts   |      8000.00 |
| checking account       | Headquarters   | Michael Smith   |       782.16 |
| checking account       | Quincy Branch  | John Blake      |      1057.75 |
| checking account       | So. NH Branch  | Theresa Markham |     67852.33 |
| checking account       | Woburn Branch  | Paula Roberts   |      3315.77 |
| money market account   | Headquarters   | Michael Smith   |     14832.64 |
| money market account   | Quincy Branch  | John Blake      |      2212.50 |
| savings account        | Headquarters   | Michael Smith   |       767.77 |
| savings account        | So. NH Branch  | Theresa Markham |       387.99 |
| savings account        | Woburn Branch  | Paula Roberts   |       700.00 |
+------------------------+----------------+-----------------+--------------+
11 rows in set (0.00 sec)
```

This query sums all deposit account balances by account type, the employee that opened the accounts, and the branches at which the accounts were opened. If you look at the query closely, you will see that the product, branch, and employee tables are needed only for display purposes, and that the account table has everything needed to generate the groupings (product_cd, open_branch_id, open_emp_id, and avail_balance). Therefore, you could separate out the task of generating the groups into a subquery, and then join the other three tables to the table generated by the subquery to achieve the desired end result. Here's the grouping subquery:

```
mysql> SELECT product_cd, open_branch_id branch_id, open_emp_id emp_id,
    ->   SUM(avail_balance) tot_deposits
    -> FROM account
    -> GROUP BY product_cd, open_branch_id, open_emp_id;
+------------+-----------+--------+--------------+
| product_cd | branch_id | emp_id | tot_deposits |
+------------+-----------+--------+--------------+
```

```
| BUS         |          2 |       10 |      9345.55 |
| BUS         |          4 |       16 |         0.00 |
| CD          |          1 |        1 |     11500.00 |
| CD          |          2 |       10 |      8000.00 |
| CHK         |          1 |        1 |       782.16 |
| CHK         |          2 |       10 |      3315.77 |
| CHK         |          3 |       13 |      1057.75 |
| CHK         |          4 |       16 |     67852.33 |
| MM          |          1 |        1 |     14832.64 |
| MM          |          3 |       13 |      2212.50 |
| SAV         |          1 |        1 |       767.77 |
| SAV         |          2 |       10 |       700.00 |
| SAV         |          4 |       16 |       387.99 |
| SBL         |          3 |       13 |     50000.00 |
+-------------+------------+----------+--------------+
14 rows in set (0.02 sec)
```

This is the heart of the query; the other tables are needed only to provide meaningful strings in place of the product_cd, open_branch_id, and open_emp_id foreign key columns. The next query wraps the query against the account table in a subquery and joins the table that results to the other three tables:

```
mysql> SELECT p.name product, b.name branch,
    ->   CONCAT(e.fname, ' ', e.lname) name,
    ->   account_groups.tot_deposits
    -> FROM
    ->  (SELECT product_cd, open_branch_id branch_id,
    ->     open_emp_id emp_id,
    ->     SUM(avail_balance) tot_deposits
    ->   FROM account
    ->   GROUP BY product_cd, open_branch_id, open_emp_id) account_groups
    ->   INNER JOIN employee e ON e.emp_id = account_groups.emp_id
    ->   INNER JOIN branch b ON b.branch_id = account_groups.branch_id
    ->   INNER JOIN product p ON p.product_cd = account_groups.product_cd
    -> WHERE p.product_type_cd = 'ACCOUNT';
+------------------------+---------------+-----------------+--------------+
| product                | branch        | name            | tot_deposits |
+------------------------+---------------+-----------------+--------------+
| certificate of deposit | Headquarters  | Michael Smith   |     11500.00 |
| certificate of deposit | Woburn Branch | Paula Roberts   |      8000.00 |
| checking account       | Headquarters  | Michael Smith   |       782.16 |
| checking account       | Quincy Branch | John Blake      |      1057.75 |
| checking account       | So. NH Branch | Theresa Markham |     67852.33 |
| checking account       | Woburn Branch | Paula Roberts   |      3315.77 |
| money market account   | Headquarters  | Michael Smith   |     14832.64 |
| money market account   | Quincy Branch | John Blake      |      2212.50 |
| savings account        | Headquarters  | Michael Smith   |       767.77 |
| savings account        | So. NH Branch | Theresa Markham |       387.99 |
| savings account        | Woburn Branch | Paula Roberts   |       700.00 |
+------------------------+---------------+-----------------+--------------+
11 rows in set (0.01 sec)
```

I realize that beauty is in the eye of the beholder, but I find this version of the query to be far more satisfying than the big, flat version. This version may execute faster, as well, because the grouping is being done on small, numeric foreign key columns (product_cd, `open_branch_id`, `open_emp_id`) instead of potentially lengthy string columns (`branch.name`, `product.name`, `employee.fname`, `employee.lname`).

Subqueries in Filter Conditions

Many of the examples in this chapter used subqueries as expressions in filter conditions, so it should not surprise you that this is one of the main uses for subqueries. However, filter conditions using subqueries are not found only in the `where` clause. For example, the next query uses a subquery in the `having` clause to find the employee responsible for opening the most accounts:

```
mysql> SELECT open_emp_id, COUNT(*) how_many
    -> FROM account
    -> GROUP BY open_emp_id
    -> HAVING COUNT(*) = (SELECT MAX(emp_cnt.how_many)
    ->   FROM (SELECT COUNT(*) how_many
    ->     FROM account
    ->     GROUP BY open_emp_id) emp_cnt);
+-------------+----------+
| open_emp_id | how_many |
+-------------+----------+
|           1 |        8 |
+-------------+----------+
1 row in set (0.01 sec)
```

The subquery in the `having` clause finds the maximum number of accounts opened by any employee, and the containing query finds the employee that has opened that number of accounts. If multiple employees tie for the highest number of opened accounts, then the query would return multiple rows.

Subqueries As Expression Generators

For this last section of the chapter, I finish where I began: with single-column, single-row scalar subqueries. Along with being used in filter conditions, scalar subqueries may be used wherever an expression can appear, including the `select` and `order by` clauses of a query and the `values` clause of an `insert` statement.

In "Task-oriented subqueries" on page 175, I showed you how to use a subquery to separate out the grouping mechanism from the rest of the query. Here's another version of the same query that uses subqueries for the same purpose, but in a different way:

```
mysql> SELECT
    -> (SELECT p.name FROM product p
    ->   WHERE p.product_cd = a.product_cd
    ->     AND p.product_type_cd = 'ACCOUNT') product,
    -> (SELECT b.name FROM branch b
    ->   WHERE b.branch_id = a.open_branch_id) branch,
    -> (SELECT CONCAT(e.fname, ' ', e.lname) FROM employee e
    ->   WHERE e.emp_id = a.open_emp_id) name,
    ->   SUM(a.avail_balance) tot_deposits
    -> FROM account a
```

```
    -> GROUP BY a.product_cd, a.open_branch_id, a.open_emp_id
    -> ORDER BY 1,2;
+------------------------+----------------+------------------+---------------+
| product                | branch         | name             | tot_deposits  |
+------------------------+----------------+------------------+---------------+
| NULL                   | Quincy Branch  | John Blake       |     50000.00  |
| NULL                   | So. NH Branch  | Theresa Markham  |         0.00  |
| NULL                   | Woburn Branch  | Paula Roberts    |      9345.55  |
| certificate of deposit | Headquarters   | Michael Smith    |     11500.00  |
| certificate of deposit | Woburn Branch  | Paula Roberts    |      8000.00  |
| checking account       | Headquarters   | Michael Smith    |       782.16  |
| checking account       | Quincy Branch  | John Blake       |      1057.75  |
| checking account       | So. NH Branch  | Theresa Markham  |     67852.33  |
| checking account       | Woburn Branch  | Paula Roberts    |      3315.77  |
| money market account   | Headquarters   | Michael Smith    |     14832.64  |
| money market account   | Quincy Branch  | John Blake       |      2212.50  |
| savings account        | Headquarters   | Michael Smith    |       767.77  |
| savings account        | So. NH Branch  | Theresa Markham  |       387.99  |
| savings account        | Woburn Branch  | Paula Roberts    |       700.00  |
+------------------------+----------------+------------------+---------------+
14 rows in set (0.01 sec)
```

There are two main differences between this query and the earlier version using a subquery in the `from` clause:

- Instead of joining the `product`, `branch`, and `employee` tables to the account data, correlated scalar subqueries are used in the `select` clause to look up the product, branch, and employee names.

- The result set has 14 rows instead of 11 rows, and three of the product names are `null`.

The reason for the extra three rows in the result set is that the previous version of the query included the filter condition `p.product_type_cd = 'ACCOUNT'`. That filter eliminated rows with product types of `INSURANCE` and `LOAN`, such as small business loans. Since this version of the query doesn't include a join to the `product` table, there is no way to include the filter condition in the main query. The correlated subquery against the `product` table does include this filter, but the only effect is to leave the product name `null`. If you want to get rid of the extra three rows, you could join the `product` table to the `account` table and include the filter condition, or you could simply do the following:

```
mysql> SELECT all_prods.product, all_prods.branch,
    ->   all_prods.name, all_prods.tot_deposits
    -> FROM
    -> (SELECT
    ->   (SELECT p.name FROM product p
    ->     WHERE p.product_cd = a.product_cd
    ->       AND p.product_type_cd = 'ACCOUNT') product,
    ->   (SELECT b.name FROM branch b
    ->     WHERE b.branch_id = a.open_branch_id) branch,
    ->   (SELECT CONCAT(e.fname, ' ', e.lname) FROM employee e
    ->     WHERE e.emp_id = a.open_emp_id) name,
    ->   SUM(a.avail_balance) tot_deposits
    ->  FROM account a
```

```
    ->   GROUP BY a.product_cd, a.open_branch_id, a.open_emp_id
    ->   ) all_prods
    -> WHERE all_prods.product IS NOT NULL
    -> ORDER BY 1,2;
+------------------------+---------------+------------------+--------------+
| product                | branch        | name             | tot_deposits |
+------------------------+---------------+------------------+--------------+
| certificate of deposit | Headquarters  | Michael Smith    |     11500.00 |
| certificate of deposit | Woburn Branch | Paula Roberts    |      8000.00 |
| checking account       | Headquarters  | Michael Smith    |       782.16 |
| checking account       | Quincy Branch | John Blake       |      1057.75 |
| checking account       | So. NH Branch | Theresa Markham  |     67852.33 |
| checking account       | Woburn Branch | Paula Roberts    |      3315.77 |
| money market account   | Headquarters  | Michael Smith    |     14832.64 |
| money market account   | Quincy Branch | John Blake       |      2212.50 |
| savings account        | Headquarters  | Michael Smith    |       767.77 |
| savings account        | So. NH Branch | Theresa Markham  |       387.99 |
| savings account        | Woburn Branch | Paula Roberts    |       700.00 |
+------------------------+---------------+------------------+--------------+
11 rows in set (0.01 sec)
```

Simply by wrapping the previous query in a subquery (called `all_prods`) and adding a filter condition to exclude `null` values of the `product` column, the query now returns the desired 11 rows. The end result is a query that performs all grouping against raw data in the `account` table, and then embellishes the output using data in three other tables, and *without doing any joins*.

As previously noted, scalar subqueries can also appear in the `order by` clause. The following query retrieves employee data sorted by the last name of each employee's boss, and then by the employee's last name:

```
mysql> SELECT emp.emp_id, CONCAT(emp.fname, ' ', emp.lname) emp_name,
    ->   (SELECT CONCAT(boss.fname, ' ', boss.lname)
    ->    FROM employee boss
    ->    WHERE boss.emp_id = emp.superior_emp_id) boss_name
    -> FROM employee emp
    -> WHERE emp.superior_emp_id IS NOT NULL
    -> ORDER BY (SELECT boss.lname FROM employee boss
    ->    WHERE boss.emp_id = emp.superior_emp_id), emp.lname;
+--------+------------------+------------------+
| emp_id | emp_name         | boss_name        |
+--------+------------------+------------------+
|     14 | Cindy Mason      | John Blake       |
|     15 | Frank Portman    | John Blake       |
|      9 | Jane Grossman    | Helen Fleming    |
|      8 | Sarah Parker     | Helen Fleming    |
|      7 | Chris Tucker     | Helen Fleming    |
|     13 | John Blake       | Susan Hawthorne  |
|      6 | Helen Fleming    | Susan Hawthorne  |
|      5 | John Gooding     | Susan Hawthorne  |
|     16 | Theresa Markham  | Susan Hawthorne  |
|     10 | Paula Roberts    | Susan Hawthorne  |
|     17 | Beth Fowler      | Theresa Markham  |
|     18 | Rick Tulman      | Theresa Markham  |
```

```
|    12 | Samantha Jameson  | Paula Roberts    |
|    11 | Thomas Ziegler    | Paula Roberts    |
|     2 | Susan Barker      | Michael Smith    |
|     3 | Robert Tyler      | Michael Smith    |
|     4 | Susan Hawthorne   | Robert Tyler     |
+-------+-------------------+------------------+
17 rows in set (0.01 sec)
```

The query uses two correlated scalar subqueries: one in the select clause to retrieve the full name of each employee's boss, and another in the order by clause to return just the last name of each employee's boss for sorting purposes.

Along with using correlated scalar subqueries in select statements, you can use non-correlated scalar subqueries to generate values for an insert statement. For example, let's say you are going to generate a new account row, and you've been given the following data:

- The product name ("savings account")
- The customer's federal ID ("555-55-5555")
- The name of the branch where the account was opened ("Quincy Branch")
- The first and last names of the teller who opened the account ("Frank Portman")

Before you can create a row in the account table, you will need to look up the key values for all of these pieces of data so that you can populate the foreign key columns in the account table. You have two choices for how to go about it: execute four queries to retrieve the primary key values and place those values into an insert statement, or use subqueries to retrieve the four key values from within an insert statement. Here's an example of the latter approach:

```sql
INSERT INTO account
  (account_id, product_cd, cust_id, open_date, last_activity_date,
   status, open_branch_id, open_emp_id, avail_balance, pending_balance)
VALUES (NULL,
  (SELECT product_cd FROM product WHERE name = 'savings account'),
  (SELECT cust_id FROM customer WHERE fed_id = '555-55-5555'),
   '2008-09-25', '2008-09-25', 'ACTIVE',
  (SELECT branch_id FROM branch WHERE name = 'Quincy Branch'),
  (SELECT emp_id FROM employee WHERE lname = 'Portman' AND fname = 'Frank'),
   0, 0);
```

Using a single SQL statement, you can create a row in the account table and look up four foreign key column values at the same time. There is one downside to this approach, however. When you use subqueries to generate data for columns that allow null values, your insert statement will succeed even if one of your subqueries fails to return a value. For example, if you mistyped Frank Portman's name in the fourth subquery, a row will still be created in account, but the open_emp_id would be set to null.

Subquery Wrap-up

I covered a lot of ground in this chapter, so it might be a good idea to review it. The examples I used in this chapter demonstrated subqueries that:

- Return a single column and row, a single column with multiple rows, and multiple columns and rows
- Are independent of the containing statement (noncorrelated subqueries)
- Reference one or more columns from the containing statement (correlated subqueries)
- Are used in conditions that utilize comparison operators as well as the special-purpose operators `in`, `not in`, `exists`, and `not exists`
- Can be found in `select`, `update`, `delete`, and `insert` statements
- Generate result sets that can be joined to other tables (or subqueries) in a query
- Can be used to generate values to populate a table or to populate columns in a query's result set
- Are used in the `select`, `from`, `where`, `having`, and `order by` clauses of queries

Obviously, subqueries are a very versatile tool, so don't feel bad if all these concepts haven't sunk in after reading this chapter for the first time. Keep experimenting with the various uses for subqueries, and you will soon find yourself thinking about how you might utilize a subquery every time you write a nontrivial SQL statement.

Test Your Knowledge

These exercises are designed to test your understanding of subqueries. Please see Appendix C for the solutions.

Exercise 9-1

Construct a query against the `account` table that uses a filter condition with a noncorrelated subquery against the `product` table to find all loan accounts (`product.product_type_cd = 'LOAN'`). Retrieve the account ID, product code, customer ID, and available balance.

Exercise 9-2

Rework the query from Exercise 9-1 using a *correlated* subquery against the `product` table to achieve the same results.

Exercise 9-3

Join the following query to the `employee` table to show the experience level of each employee:

```
SELECT 'trainee' name, '2004-01-01' start_dt, '2005-12-31' end_dt
UNION ALL
SELECT 'worker' name, '2002-01-01' start_dt, '2003-12-31' end_dt
UNION ALL
SELECT 'mentor' name, '2000-01-01' start_dt, '2001-12-31' end_dt
```

Give the subquery the alias `levels`, and include the employee ID, first name, last name, and experience level (`levels.name`). (Hint: build a join condition using an inequality condition to determine into which level the `employee.start_date` column falls.)

Exercise 9-4

Construct a query against the `employee` table that retrieves the employee ID, first name, and last name, along with the names of the department and branch to which the employee is assigned. Do not join any tables.

CHAPTER 10

Joins Revisited

By now, you should be comfortable with the concept of the inner join, which I introduced in Chapter 5. This chapter focuses on other ways in which you can join tables, including the outer join and the cross join.

Outer Joins

In all the examples thus far that have included multiple tables, we haven't been concerned that the join conditions might fail to find matches for all the rows in the tables. For example, when joining the `account` table to the `customer` table, I did not mention the possibility that a value in the `cust_id` column of the `account` table might not match a value in the `cust_id` column of the `customer` table. If that were the case, then some of the rows in one table or the other would be left out of the result set.

Just to be sure, let's check the data in the tables. Here are the `account_id` and `cust_id` columns from the `account` table:

```
mysql> SELECT account_id, cust_id
    -> FROM account;
+------------+---------+
| account_id | cust_id |
+------------+---------+
|          1 |       1 |
|          2 |       1 |
|          3 |       1 |
|          4 |       2 |
|          5 |       2 |
|          7 |       3 |
|          8 |       3 |
|         10 |       4 |
|         11 |       4 |
|         12 |       4 |
|         13 |       5 |
|         14 |       6 |
|         15 |       6 |
|         17 |       7 |
|         18 |       8 |
```

```
|        19 |       8 |
|        21 |       9 |
|        22 |       9 |
|        23 |       9 |
|        24 |      10 |
|        25 |      10 |
|        27 |      11 |
|        28 |      12 |
|        29 |      13 |
+-----------+---------+
24 rows in set (1.50 sec)
```

There are 24 accounts spanning 13 different customers, with customer IDs 1 through 13 having at least one account. Here's the set of customer IDs from the customer table:

```
mysql> SELECT cust_id
    -> FROM customer;

+---------+
| cust_id |
+---------+
|       1 |
|       2 |
|       3 |
|       4 |
|       5 |
|       6 |
|       7 |
|       8 |
|       9 |
|      10 |
|      11 |
|      12 |
|      13 |
+---------+
13 rows in set (0.02 sec)
```

There are 13 rows in the customer table with IDs 1 through 13, so every customer ID is included at least once in the account table. When the two tables are joined on the cust_id column, therefore, you would expect all 24 rows to be included in the result set (barring any other filter conditions):

```
mysql> SELECT a.account_id, c.cust_id
    -> FROM account a INNER JOIN customer c
    ->   ON a.cust_id = c.cust_id;
+------------+---------+
| account_id | cust_id |
+------------+---------+
|          1 |       1 |
|          2 |       1 |
|          3 |       1 |
|          4 |       2 |
|          5 |       2 |
|          7 |       3 |
|          8 |       3 |
```

```
|     10 |        4 |
|     11 |        4 |
|     12 |        4 |
|     13 |        5 |
|     14 |        6 |
|     15 |        6 |
|     17 |        7 |
|     18 |        8 |
|     19 |        8 |
|     21 |        9 |
|     22 |        9 |
|     23 |        9 |
|     24 |       10 |
|     25 |       10 |
|     27 |       11 |
|     28 |       12 |
|     29 |       13 |
+-----------+---------+
24 rows in set (0.06 sec)
```

As expected, all 24 accounts are present in the result set. But what happens if you join the account table to one of the specialized customer tables, such as the business table?

```
mysql> SELECT a.account_id, b.cust_id, b.name
    -> FROM account a INNER JOIN business b
    ->   ON a.cust_id = b.cust_id;
+------------+---------+-------------------------+
| account_id | cust_id | name                    |
+------------+---------+-------------------------+
|         24 |      10 | Chilton Engineering     |
|         25 |      10 | Chilton Engineering     |
|         27 |      11 | Northeast Cooling Inc.  |
|         28 |      12 | Superior Auto Body      |
|         29 |      13 | AAA Insurance Inc.      |
+------------+---------+-------------------------+
5 rows in set (0.10 sec)
```

Instead of 24 rows in the result set, there are now only five. Let's look in the business table to see why this is:

```
mysql> SELECT cust_id, name
    -> FROM business;
+---------+-------------------------+
| cust_id | name                    |
+---------+-------------------------+
|      10 | Chilton Engineering     |
|      11 | Northeast Cooling Inc.  |
|      12 | Superior Auto Body      |
|      13 | AAA Insurance Inc.      |
+---------+-------------------------+
4 rows in set (0.01 sec)
```

Of the 13 rows in the customer table, only four are business customers, and since one of the business customers has two accounts, a total of five rows in the account table are linked to business customers.

But what if you want your query to return *all* the accounts, but to include the business name only if the account is linked to a business customer? This is an example where you would need an *outer join* between the account and business tables, as in:

```
mysql> SELECT a.account_id, a.cust_id, b.name
    -> FROM account a LEFT OUTER JOIN business b
    ->   ON a.cust_id = b.cust_id;
+------------+---------+------------------------+
| account_id | cust_id | name                   |
+------------+---------+------------------------+
|          1 |       1 | NULL                   |
|          2 |       1 | NULL                   |
|          3 |       1 | NULL                   |
|          4 |       2 | NULL                   |
|          5 |       2 | NULL                   |
|          7 |       3 | NULL                   |
|          8 |       3 | NULL                   |
|         10 |       4 | NULL                   |
|         11 |       4 | NULL                   |
|         12 |       4 | NULL                   |
|         13 |       5 | NULL                   |
|         14 |       6 | NULL                   |
|         15 |       6 | NULL                   |
|         17 |       7 | NULL                   |
|         18 |       8 | NULL                   |
|         19 |       8 | NULL                   |
|         21 |       9 | NULL                   |
|         22 |       9 | NULL                   |
|         23 |       9 | NULL                   |
|         24 |      10 | Chilton Engineering    |
|         25 |      10 | Chilton Engineering    |
|         27 |      11 | Northeast Cooling Inc. |
|         28 |      12 |  Superior Auto Body    |
|         29 |      13 | AAA Insurance Inc.     |
+------------+---------+------------------------+
24 rows in set (0.04 sec)
```

An outer join includes all of the rows from one table and includes data from the second table only if matching rows are found. In this case, all rows from the account table are included, since I specified left outer join and the account table is on the left side of the join definition. The name column is null for all rows except for the four business customers (cust_ids 10, 11, 12, and 13). Here's a similar query with an outer join to the individual table instead of the business table:

```
mysql> SELECT a.account_id, a.cust_id, i.fname, i.lname
    -> FROM account a LEFT OUTER JOIN individual i
    ->   ON a.cust_id = i.cust_id;
+------------+---------+---------+---------+
| account_id | cust_id | fname   | lname   |
+------------+---------+---------+---------+
|          1 |       1 | James   | Hadley  |
|          2 |       1 | James   | Hadley  |
|          3 |       1 | James   | Hadley  |
|          4 |       2 | Susan   | Tingley |
```

```
|         5 |        2 | Susan    | Tingley  |
|         7 |        3 | Frank    | Tucker   |
|         8 |        3 | Frank    | Tucker   |
|        10 |        4 | John     | Hayward  |
|        11 |        4 | John     | Hayward  |
|        12 |        4 | John     | Hayward  |
|        13 |        5 | Charles  | Frasier  |
|        14 |        6 | John     | Spencer  |
|        15 |        6 | John     | Spencer  |
|        17 |        7 | Margaret | Young    |
|        18 |        8 | George   | Blake    |
|        19 |        8 | George   | Blake    |
|        21 |        9 | Richard  | Farley   |
|        22 |        9 | Richard  | Farley   |
|        23 |        9 | Richard  | Farley   |
|        24 |       10 | NULL     | NULL     |
|        25 |       10 | NULL     | NULL     |
|        27 |       11 | NULL     | NULL     |
|        28 |       12 | NULL     | NULL     |
|        29 |       13 | NULL     | NULL     |
+-----------+----------+----------+----------+
24 rows in set (0.09 sec)
```

This query is essentially the reverse of the previous query: first and last names are supplied for the individual customers, whereas the columns are `null` for the business customers.

Left Versus Right Outer Joins

In each of the outer join examples in the previous section, I specified `left outer join`. The keyword `left` indicates that the table on the left side of the `join` is responsible for determining the number of rows in the result set, whereas the table on the right side is used to provide column values whenever a match is found. Consider the following query:

```
mysql> SELECT c.cust_id, b.name
    -> FROM customer c LEFT OUTER JOIN business b
    ->   ON c.cust_id = b.cust_id;
+---------+------------------------+
| cust_id | name                   |
+---------+------------------------+
|       1 | NULL                   |
|       2 | NULL                   |
|       3 | NULL                   |
|       4 | NULL                   |
|       5 | NULL                   |
|       6 | NULL                   |
|       7 | NULL                   |
|       8 | NULL                   |
|       9 | NULL                   |
|      10 | Chilton Engineering    |
|      11 | Northeast Cooling Inc. |
|      12 | Superior Auto Body     |
```

```
|       13 | AAA Insurance Inc.   |
+----------+-----------------------+
13 rows in set (0.00 sec)
```

The `from` clause specifies a left outer join, so all 13 rows from the `customer` table are included in the result set, with the `business` table contributing values to the second column in the result set for the four business customers. If you execute the same query, but indicate `right outer join`, you would see the following results:

```
mysql> SELECT c.cust_id, b.name
    -> FROM customer c RIGHT OUTER JOIN business b
    ->   ON c.cust_id = b.cust_id;
+----------+-----------------------+
| cust_id | name                   |
+----------+-----------------------+
|       10 | Chilton Engineering   |
|       11 | Northeast Cooling Inc. |
|       12 | Superior Auto Body    |
|       13 | AAA Insurance Inc.    |
+----------+-----------------------+
4 rows in set (0.00 sec)
```

The number of rows in the result set is now determined by the number of rows in the `business` table, which is why there are only four rows in the result set.

Keep in mind that both queries are performing outer joins; the keywords `left` and `right` are there just to tell the server which table is allowed to have gaps in the data. If you want to outer-join tables A and B and you want all rows from A with additional columns from B whenever there is matching data, you can specify either A `left outer join` B or B `right outer join` A.

Three-Way Outer Joins

In some cases, you may want to outer-join one table with two other tables. For example, you may want a list of all accounts showing either the customer's first and last names for individuals or the business name for business customers, as in:

```
mysql> SELECT a.account_id, a.product_cd,
    ->   CONCAT(i.fname, ' ', i.lname) person_name,
    ->   b.name business_name
    -> FROM account a LEFT OUTER JOIN individual i
    ->   ON a.cust_id = i.cust_id
    ->   LEFT OUTER JOIN business b
    ->   ON a.cust_id = b.cust_id;
+------------+------------+-----------------+-----------------------+
| account_id | product_cd | person_name     | business_name         |
+------------+------------+-----------------+-----------------------+
|          1 | CHK        | James Hadley    | NULL                  |
|          2 | SAV        | James Hadley    | NULL                  |
|          3 | CD         | James Hadley    | NULL                  |
|          4 | CHK        | Susan Tingley   | NULL                  |
|          5 | SAV        | Susan Tingley   | NULL                  |
|          7 | CHK        | Frank Tucker    | NULL                  |
```

```
|         8 | MM         | Frank Tucker     | NULL                  |
|        10 | CHK        | John Hayward     | NULL                  |
|        11 | SAV        | John Hayward     | NULL                  |
|        12 | MM         | John Hayward     | NULL                  |
|        13 | CHK        | Charles Frasier  | NULL                  |
|        14 | CHK        | John Spencer     | NULL                  |
|        15 | CD         | John Spencer     | NULL                  |
|        17 | CD         | Margaret Young   | NULL                  |
|        18 | CHK        | George Blake     | NULL                  |
|        19 | SAV        | George Blake     | NULL                  |
|        21 | CHK        | Richard Farley   | NULL                  |
|        22 | MM         | Richard Farley   | NULL                  |
|        23 | CD         | Richard Farley   | NULL                  |
|        24 | CHK        | NULL             | Chilton Engineering   |
|        25 | BUS        | NULL             | Chilton Engineering   |
|        27 | BUS        | NULL             | Northeast Cooling Inc.|
|        28 | CHK        | NULL             | Superior Auto Body    |
|        29 | SBL        | NULL             | AAA Insurance Inc.    |
+-----------+------------+------------------+-----------------------+
24 rows in set (0.08 sec)
```

The results include all 24 rows from the **account** table, along with either a person's name or a business name coming from the two outer-joined tables.

I don't know of any restrictions with MySQL regarding the number of tables that can be outer-joined to the same table, but you can always use subqueries to limit the number of joins in your query. For instance, you can rewrite the previous example as follows:

```
mysql> SELECT account_ind.account_id, account_ind.product_cd,
    ->   account_ind.person_name,
    ->   b.name business_name
    -> FROM
    -> (SELECT a.account_id, a.product_cd, a.cust_id,
    ->    CONCAT(i.fname, ' ', i.lname) person_name
    ->  FROM account a LEFT OUTER JOIN individual i
    ->    ON a.cust_id = i.cust_id) account_ind
    ->  LEFT OUTER JOIN business b
    ->    ON account_ind.cust_id = b.cust_id;
+------------+------------+------------------+-----------------------+
| account_id | product_cd | person_name      | business_name         |
+------------+------------+------------------+-----------------------+
|          1 | CHK        | James Hadley     | NULL                  |
|          2 | SAV        | James Hadley     | NULL                  |
|          3 | CD         | James Hadley     | NULL                  |
|          4 | CHK        | Susan Tingley    | NULL                  |
|          5 | SAV        | Susan Tingley    | NULL                  |
|          7 | CHK        | Frank Tucker     | NULL                  |
|          8 | MM         | Frank Tucker     | NULL                  |
|         10 | CHK        | John Hayward     | NULL                  |
|         11 | SAV        | John Hayward     | NULL                  |
|         12 | MM         | John Hayward     | NULL                  |
|         13 | CHK        | Charles Frasier  | NULL                  |
|         14 | CHK        | John Spencer     | NULL                  |
|         15 | CD         | John Spencer     | NULL                  |
|         17 | CD         | Margaret Young   | NULL                  |
```

```
|        18 | CHK        | George Blake     | NULL                   |
|        19 | SAV        | George Blake     | NULL                   |
|        21 | CHK        | Richard Farley   | NULL                   |
|        22 | MM         | Richard Farley   | NULL                   |
|        23 | CD         | Richard Farley   | NULL                   |
|        24 | CHK        | NULL             | Chilton Engineering    |
|        25 | BUS        | NULL             | Chilton Engineering    |
|        27 | BUS        | NULL             | Northeast Cooling Inc. |
|        28 | CHK        | NULL             | Superior Auto Body     |
|        29 | SBL        | NULL             | AAA Insurance Inc.     |
+-----------+------------+------------------+------------------------+
24 rows in set (0.08 sec)
```

In this version of the query, the `individual` table is outer-joined to the `account` table within a subquery named `account_ind`, the results of which are then outer-joined to the `business` table. Thus, each query (the subquery and the containing query) uses only a single outer join. If you are using a database other than MySQL, you may need to utilize this strategy if you want to outer-join more than one table.

Self Outer Joins

In Chapter 5, I introduced you to the concept of the self-join, where a table is joined to itself. Here's a self-join example from Chapter 5, which joins the `employee` table to itself to generate a list of employees and their supervisors:

```
mysql> SELECT e.fname, e.lname,
    ->   e_mgr.fname mgr_fname, e_mgr.lname mgr_lname
    -> FROM employee e INNER JOIN employee e_mgr
    ->   ON e.superior_emp_id = e_mgr.emp_id;
+----------+-----------+-----------+-----------+
| fname    | lname     | mgr_fname | mgr_lname |
+----------+-----------+-----------+-----------+
| Susan    | Barker    | Michael   | Smith     |
| Robert   | Tyler     | Michael   | Smith     |
| Susan    | Hawthorne | Robert    | Tyler     |
| John     | Gooding   | Susan     | Hawthorne |
| Helen    | Fleming   | Susan     | Hawthorne |
| Chris    | Tucker    | Helen     | Fleming   |
| Sarah    | Parker    | Helen     | Fleming   |
| Jane     | Grossman  | Helen     | Fleming   |
| Paula    | Roberts   | Susan     | Hawthorne |
| Thomas   | Ziegler   | Paula     | Roberts   |
| Samantha | Jameson   | Paula     | Roberts   |
| John     | Blake     | Susan     | Hawthorne |
| Cindy    | Mason     | John      | Blake     |
| Frank    | Portman   | John      | Blake     |
| Theresa  | Markham   | Susan     | Hawthorne |
| Beth     | Fowler    | Theresa   | Markham   |
| Rick     | Tulman    | Theresa   | Markham   |
+----------+-----------+-----------+-----------+
17 rows in set (0.02 sec)
```

This query works fine except for one small issue: employees who don't have a supervisor are left out of the result set. By changing the join from an inner join to an outer join, however, the result set will include all employees, including those without supervisors:

```
mysql> SELECT e.fname, e.lname,
    ->   e_mgr.fname mgr_fname, e_mgr.lname mgr_lname
    -> FROM employee e LEFT OUTER JOIN employee e_mgr
    ->   ON e.superior_emp_id = e_mgr.emp_id;
+----------+-----------+-----------+-----------+
| fname    | lname     | mgr_fname | mgr_lname |
+----------+-----------+-----------+-----------+
| Michael  | Smith     | NULL      | NULL      |
| Susan    | Barker    | Michael   | Smith     |
| Robert   | Tyler     | Michael   | Smith     |
| Susan    | Hawthorne | Robert    | Tyler     |
| John     | Gooding   | Susan     | Hawthorne |
| Helen    | Fleming   | Susan     | Hawthorne |
| Chris    | Tucker    | Helen     | Fleming   |
| Sarah    | Parker    | Helen     | Fleming   |
| Jane     | Grossman  | Helen     | Fleming   |
| Paula    | Roberts   | Susan     | Hawthorne |
| Thomas   | Ziegler   | Paula     | Roberts   |
| Samantha | Jameson   | Paula     | Roberts   |
| John     | Blake     | Susan     | Hawthorne |
| Cindy    | Mason     | John      | Blake     |
| Frank    | Portman   | John      | Blake     |
| Theresa  | Markham   | Susan     | Hawthorne |
| Beth     | Fowler    | Theresa   | Markham   |
| Rick     | Tulman    | Theresa   | Markham   |
+----------+-----------+-----------+-----------+
18 rows in set (0.00 sec)
```

The result set now includes Michael Smith, who is the president of the bank and, therefore, does not have a supervisor. The query utilizes a left outer join to generate a list of all employees and, if applicable, their supervisor. If you change the join to be a right outer join, you would see the following results:

```
mysql> SELECT e.fname, e.lname,
    ->   e_mgr.fname mgr_fname, e_mgr.lname mgr_lname
    -> FROM employee e RIGHT OUTER JOIN employee e_mgr
    ->   ON e.superior_emp_id = e_mgr.emp_id;
+----------+-----------+-----------+-----------+
| fname    | lname     | mgr_fname | mgr_lname |
+----------+-----------+-----------+-----------+
| Susan    | Barker    | Michael   | Smith     |
| Robert   | Tyler     | Michael   | Smith     |
| NULL     | NULL      | Susan     | Barker    |
| Susan    | Hawthorne | Robert    | Tyler     |
| John     | Gooding   | Susan     | Hawthorne |
| Helen    | Fleming   | Susan     | Hawthorne |
| Paula    | Roberts   | Susan     | Hawthorne |
| John     | Blake     | Susan     | Hawthorne |
| Theresa  | Markham   | Susan     | Hawthorne |
| NULL     | NULL      | John      | Gooding   |
| Chris    | Tucker    | Helen     | Fleming   |
```

```
| Sarah    | Parker    | Helen    | Fleming  |
| Jane     | Grossman  | Helen    | Fleming  |
| NULL     | NULL      | Chris    | Tucker   |
| NULL     | NULL      | Sarah    | Parker   |
| NULL     | NULL      | Jane     | Grossman |
| Thomas   | Ziegler   | Paula    | Roberts  |
| Samantha | Jameson   | Paula    | Roberts  |
| NULL     | NULL      | Thomas   | Ziegler  |
| NULL     | NULL      | Samantha | Jameson  |
| Cindy    | Mason     | John     | Blake    |
| Frank    | Portman   | John     | Blake    |
| NULL     | NULL      | Cindy    | Mason    |
| NULL     | NULL      | Frank    | Portman  |
| Beth     | Fowler    | Theresa  | Markham  |
| Rick     | Tulman    | Theresa  | Markham  |
| NULL     | NULL      | Beth     | Fowler   |
| NULL     | NULL      | Rick     | Tulman   |
+----------+-----------+----------+----------+
28 rows in set (0.00 sec)
```

This query shows each supervisor (still the third and fourth columns) along with the set of employees he or she supervises. Therefore, Michael Smith appears twice as supervisor to Susan Barker and Robert Tyler; Susan Barker appears once as a supervisor to nobody (null values in the first and second columns). All 18 employees appear at least once in the third and fourth columns, with some appearing more than once if they supervise more than one employee, making a total of 28 rows in the result set. This is a very different outcome from the previous query, and it was prompted by changing only a single keyword (left to right). Therefore, when using outer joins, make sure you think carefully about whether to specify a left or right outer join.

Cross Joins

Back in Chapter 5, I introduced the concept of a Cartesian product, which is essentially the result of joining multiple tables without specifying any join conditions. Cartesian products are used fairly frequently by accident (e.g., forgetting to add the join condition to the from clause) but are not so common otherwise. If, however, you *do* intend to generate the Cartesian product of two tables, you should specify a *cross join*, as in:

```
mysql> SELECT pt.name, p.product_cd, p.name
    -> FROM product p CROSS JOIN product_type pt;
+-------------------+------------+------------------------+
| name              | product_cd | name                   |
+-------------------+------------+------------------------+
| Customer Accounts | AUT        | auto loan              |
| Customer Accounts | BUS        | business line of credit |
| Customer Accounts | CD         | certificate of deposit |
| Customer Accounts | CHK        | checking account       |
| Customer Accounts | MM         | money market account   |
| Customer Accounts | MRT        | home mortgage          |
| Customer Accounts | SAV        | savings account        |
| Customer Accounts | SBL        | small business loan    |
```

```
| Insurance Offerings          | AUT | auto loan                |
| Insurance Offerings          | BUS | business line of credit  |
| Insurance Offerings          | CD  | certificate of deposit   |
| Insurance Offerings          | CHK | checking account         |
| Insurance Offerings          | MM  | money market account     |
| Insurance Offerings          | MRT | home mortgage            |
| Insurance Offerings          | SAV | savings account          |
| Insurance Offerings          | SBL | small business loan      |
| Individual and Business Loans | AUT | auto loan               |
| Individual and Business Loans | BUS | business line of credit |
| Individual and Business Loans | CD  | certificate of deposit  |
| Individual and Business Loans | CHK | checking account        |
| Individual and Business Loans | MM  | money market account    |
| Individual and Business Loans | MRT | home mortgage           |
| Individual and Business Loans | SAV | savings account         |
| Individual and Business Loans | SBL | small business loan     |
+-------------------------------+-----------+-------------------------+
24 rows in set (0.00 sec)
```

This query generates the Cartesian product of the `product` and `product_type` tables, resulting in 24 rows (8 `product` rows × 3 `product_type` rows). But now that you know what a cross join is and how to specify it, what is it used for? Most SQL books will describe what a cross join is and then tell you that it is seldom useful, but I would like to share with you a situation in which I find the cross join to be quite helpful.

In Chapter 9, I discussed how to use subqueries to fabricate tables. The example I used showed how to build a three-row table that could be joined to other tables. Here's the fabricated table from the example:

```
mysql> SELECT 'Small Fry' name, 0 low_limit, 4999.99 high_limit
    -> UNION ALL
    -> SELECT 'Average Joes' name, 5000 low_limit, 9999.99 high_limit
    -> UNION ALL
    -> SELECT 'Heavy Hitters' name, 10000 low_limit, 9999999.99 high_limit;
+---------------+-----------+------------+
| name          | low_limit | high_limit |
+---------------+-----------+------------+
| Small Fry     |         0 |    4999.99 |
| Average Joes  |      5000 |    9999.99 |
| Heavy Hitters |     10000 | 9999999.99 |
+---------------+-----------+------------+
3 rows in set (0.00 sec)
```

While this table was exactly what was needed for placing customers into three groups based on their aggregate account balance, this strategy of merging single-row tables using the set operator `union all` doesn't work very well if you need to fabricate a large table.

Say, for example, that you want to create a query that generates a row for every day in the year 2008, but you don't have a table in your database that contains a row for every day. Using the strategy from the example in Chapter 9, you could do something like the following:

```
SELECT '2008-01-01' dt
UNION ALL
SELECT '2008-01-02' dt
UNION ALL
SELECT '2008-01-03' dt
UNION ALL
...
...
...
SELECT '2008-12-29' dt
UNION ALL
SELECT '2008-12-30' dt
UNION ALL
SELECT '2008-12-31' dt
```

Building a query that merges together the results of 366 queries is a bit tedious, so maybe a different strategy is needed. What if you generate a table with 366 rows (2008 was a leap year) with a single column containing a number between 0 and 366, and then add that number of days to January 1, 2008? Here's one possible method to generate such a table:

```
mysql> SELECT ones.num + tens.num + hundreds.num
    -> FROM
    ->  (SELECT 0 num UNION ALL
    ->    SELECT 1 num UNION ALL
    ->    SELECT 2 num UNION ALL
    ->    SELECT 3 num UNION ALL
    ->    SELECT 4 num UNION ALL
    ->    SELECT 5 num UNION ALL
    ->    SELECT 6 num UNION ALL
    ->    SELECT 7 num UNION ALL
    ->    SELECT 8 num UNION ALL
    ->    SELECT 9 num) ones
    ->  CROSS JOIN
    ->  (SELECT 0 num UNION ALL
    ->    SELECT 10 num UNION ALL
    ->    SELECT 20 num UNION ALL
    ->    SELECT 30 num UNION ALL
    ->    SELECT 40 num UNION ALL
    ->    SELECT 50 num UNION ALL
    ->    SELECT 60 num UNION ALL
    ->    SELECT 70 num UNION ALL
    ->    SELECT 80 num UNION ALL
    ->    SELECT 90 num) tens
    ->  CROSS JOIN
    ->  (SELECT 0 num UNION ALL
    ->    SELECT 100 num UNION ALL
    ->    SELECT 200 num UNION ALL
    ->    SELECT 300 num) hundreds;
+------------------------------------+
| ones.num + tens.num + hundreds.num |
+------------------------------------+
|                                  0 |
|                                  1 |
|                                  2 |
```

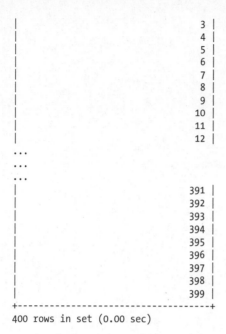

```
|                                      3 |
|                                      4 |
|                                      5 |
|                                      6 |
|                                      7 |
|                                      8 |
|                                      9 |
|                                     10 |
|                                     11 |
|                                     12 |
...
...
...
|                                    391 |
|                                    392 |
|                                    393 |
|                                    394 |
|                                    395 |
|                                    396 |
|                                    397 |
|                                    398 |
|                                    399 |
+----------------------------------------+
400 rows in set (0.00 sec)
```

If you take the Cartesian product of the three sets {0, 1, 2, 3, 4, 5, 6, 7, 8, 9}, {0, 10, 20, 30, 40, 50, 60, 70, 80, 90}, and {0, 100, 200, 300} and add the values in the three columns, you get a 400-row result set containing all numbers between 0 and 399. While this is more than the 366 rows needed to generate the set of days in 2008, it's easy enough to get rid of the excess rows, and I'll show you how shortly.

The next step is to convert the set of numbers to a set of dates. To do this, I will use the date_add() function to add each number in the result set to January 1, 2008. Then I'll add a filter condition to throw away any dates that venture into 2009:

```
mysql> SELECT DATE_ADD('2008-01-01',
    ->    INTERVAL (ones.num + tens.num + hundreds.num) DAY) dt
    -> FROM
    -> (SELECT 0 num UNION ALL
    ->    SELECT 1 num UNION ALL
    ->    SELECT 2 num UNION ALL
    ->    SELECT 3 num UNION ALL
    ->    SELECT 4 num UNION ALL
    ->    SELECT 5 num UNION ALL
    ->    SELECT 6 num UNION ALL
    ->    SELECT 7 num UNION ALL
    ->    SELECT 8 num UNION ALL
    ->    SELECT 9 num) ones
    ->  CROSS JOIN
    -> (SELECT 0 num UNION ALL
    ->    SELECT 10 num UNION ALL
    ->    SELECT 20 num UNION ALL
    ->    SELECT 30 num UNION ALL
    ->    SELECT 40 num UNION ALL
```

```
    ->     SELECT 50 num UNION ALL
    ->     SELECT 60 num UNION ALL
    ->     SELECT 70 num UNION ALL
    ->     SELECT 80 num UNION ALL
    ->     SELECT 90 num) tens
    ->   CROSS JOIN
    ->   (SELECT 0 num UNION ALL
    ->     SELECT 100 num UNION ALL
    ->     SELECT 200 num UNION ALL
    ->     SELECT 300 num) hundreds
    -> WHERE DATE_ADD('2008-01-01',
    ->   INTERVAL (ones.num + tens.num + hundreds.num) DAY) < '2009-01-01'
    -> ORDER BY 1;
+------------+
| dt         |
+------------+
| 2008-01-01 |
| 2008-01-02 |
| 2008-01-03 |
| 2008-01-04 |
| 2008-01-05 |
| 2008-01-06 |
| 2008-01-07 |
| 2008-01-08 |
| 2008-01-09 |
| 2008-01-10 |
...
...
...
| 2008-02-20 |
| 2008-02-21 |
| 2008-02-22 |
| 2008-02-23 |
| 2008-02-24 |
| 2008-02-25 |
| 2008-02-26 |
| 2008-02-27 |
| 2008-02-28 |
| 2008-02-29 |
| 2008-03-01 |
...
...
...
| 2008-12-20 |
| 2008-12-21 |
| 2008-12-22 |
| 2008-12-23 |
| 2008-12-24 |
| 2008-12-25 |
| 2008-12-26 |
| 2008-12-27 |
| 2008-12-28 |
| 2008-12-29 |
| 2008-12-30 |
| 2008-12-31 |
```

```
+------------+
366 rows in set (0.01 sec)
```

The nice thing about this approach is that the result set automatically includes the extra leap day (February 29) without your intervention, since the database server figures it out when it adds 59 days to January 1, 2008.

Now that you have a mechanism for fabricating all the days in 2008, what should you do with it? Well, you might be asked to generate a query that shows every day in 2008 along with the number of banking transactions conducted on that day, the number of accounts opened on that day, and so forth. Here's an example that answers the first question:

```
mysql> SELECT days.dt, COUNT(t.txn_id)
    -> FROM transaction t RIGHT OUTER JOIN
    ->  (SELECT DATE_ADD('2008-01-01',
    ->     INTERVAL (ones.num + tens.num + hundreds.num) DAY) dt
    ->   FROM
    ->    (SELECT 0 num UNION ALL
    ->     SELECT 1 num UNION ALL
    ->     SELECT 2 num UNION ALL
    ->     SELECT 3 num UNION ALL
    ->     SELECT 4 num UNION ALL
    ->     SELECT 5 num UNION ALL
    ->     SELECT 6 num UNION ALL
    ->     SELECT 7 num UNION ALL
    ->     SELECT 8 num UNION ALL
    ->     SELECT 9 num) ones
    ->     CROSS JOIN
    ->    (SELECT 0 num UNION ALL
    ->     SELECT 10 num UNION ALL
    ->     SELECT 20 num UNION ALL
    ->     SELECT 30 num UNION ALL
    ->     SELECT 40 num UNION ALL
    ->     SELECT 50 num UNION ALL
    ->     SELECT 60 num UNION ALL
    ->     SELECT 70 num UNION ALL
    ->     SELECT 80 num UNION ALL
    ->     SELECT 90 num) tens
    ->     CROSS JOIN
    ->    (SELECT 0 num UNION ALL
    ->     SELECT 100 num UNION ALL
    ->     SELECT 200 num UNION ALL
    ->     SELECT 300 num) hundreds
    ->   WHERE DATE_ADD('2008-01-01',
    ->     INTERVAL (ones.num + tens.num + hundreds.num) DAY) <
    ->       '2009-01-01') days
    ->   ON days.dt = t.txn_date
    -> GROUP BY days.dt
    -> ORDER BY 1;
+------------+-----------------+
| dt         | COUNT(t.txn_id) |
+------------+-----------------+
| 2008-01-01 |               0 |
```

```
| 2008-01-02 |                0 |
| 2008-01-03 |                0 |
| 2008-01-04 |                0 |
| 2008-01-05 |               21 |
| 2008-01-06 |                0 |
| 2008-01-07 |                0 |
| 2008-01-08 |                0 |
| 2008-01-09 |                0 |
| 2008-01-10 |                0 |
| 2008-01-11 |                0 |
| 2008-01-12 |                0 |
| 2008-01-13 |                0 |
| 2008-01-14 |                0 |
| 2008-01-15 |                0 |
...
| 2008-12-31 |                0 |
+------------+------------------+
366 rows in set (0.03 sec)
```

This is one of the more interesting queries thus far in the book, in that it includes cross joins, outer joins, a date function, grouping, set operations (union all), and an aggregate function (count()). It is also not the most elegant solution to the given problem, but it should serve as an example of how, with a little creativity and a firm grasp on the language, you can make even a seldom-used feature like cross joins a potent tool in your SQL toolkit.

Natural Joins

If you are lazy (and aren't we all), you can choose a join type that allows you to name the tables to be joined but lets the database server determine what the join conditions need to be. Known as the *natural join*, this join type relies on identical column names across multiple tables to infer the proper join conditions. For example, the account table includes a column named cust_id, which is the foreign key to the customer table, whose primary key is also named cust_id. Thus, you can write a query that uses natural join to join the two tables:

```
mysql> SELECT a.account_id, a.cust_id, c.cust_type_cd, c.fed_id
    -> FROM account a NATURAL JOIN customer c;
+------------+---------+--------------+-------------+
| account_id | cust_id | cust_type_cd | fed_id      |
+------------+---------+--------------+-------------+
|          1 |       1 | I            | 111-11-1111 |
|          2 |       1 | I            | 111-11-1111 |
|          3 |       1 | I            | 111-11-1111 |
|          4 |       2 | I            | 222-22-2222 |
|          5 |       2 | I            | 222-22-2222 |
|          6 |       3 | I            | 333-33-3333 |
|          7 |       3 | I            | 333-33-3333 |
|          8 |       4 | I            | 444-44-4444 |
|          9 |       4 | I            | 444-44-4444 |
|         10 |       4 | I            | 444-44-4444 |
```

```
|          11 |       5 | I          | 555-55-5555 |
|          12 |       6 | I          | 666-66-6666 |
|          13 |       6 | I          | 666-66-6666 |
|          14 |       7 | I          | 777-77-7777 |
|          15 |       8 | I          | 888-88-8888 |
|          16 |       8 | I          | 888-88-8888 |
|          17 |       9 | I          | 999-99-9999 |
|          18 |       9 | I          | 999-99-9999 |
|          19 |       9 | I          | 999-99-9999 |
|          20 |      10 | B          | 04-1111111  |
|          21 |      10 | B          | 04-1111111  |
|          22 |      11 | B          | 04-2222222  |
|          23 |      12 | B          | 04-3333333  |
|          24 |      13 | B          | 04-4444444  |
+-------------+---------+------------+-------------+
24 rows in set (0.02 sec)
```

Because you specified a natural join, the server inspected the table definitions and added the join condition a.cust_id = c.cust_id to join the two tables.

This is all well and good, but what if the columns don't have the same name across the tables? For example, the account table also has a foreign key to the branch table, but the column in the account table is named open_branch_id instead of just branch_id. Let's see what happens if I use natural join between the account and branch tables:

```
mysql> SELECT a.account_id, a.cust_id, a.open_branch_id, b.name
    -> FROM account a NATURAL JOIN branch b;
+------------+---------+----------------+---------------+
| account_id | cust_id | open_branch_id | name          |
+------------+---------+----------------+---------------+
|          1 |       1 |              2 | Headquarters  |
|          1 |       1 |              2 | Woburn Branch |
|          1 |       1 |              2 | Quincy Branch |
|          1 |       1 |              2 | So. NH Branch |
|          2 |       1 |              2 | Headquarters  |
|          2 |       1 |              2 | Woburn Branch |
|          2 |       1 |              2 | Quincy Branch |
|          2 |       1 |              2 | So. NH Branch |
|          3 |       1 |              2 | Headquarters  |
|          3 |       1 |              2 | Woburn Branch |
|          3 |       1 |              2 | Quincy Branch |
|          3 |       1 |              2 | So. NH Branch |
|          4 |       2 |              2 | Headquarters  |
|          4 |       2 |              2 | Woburn Branch |
|          4 |       2 |              2 | Quincy Branch |
|          4 |       2 |              2 | So. NH Branch |
|          5 |       2 |              2 | Headquarters  |
|          5 |       2 |              2 | Woburn Branch |
|          5 |       2 |              2 | Quincy Branch |
|          5 |       2 |              2 | So. NH Branch |
|          7 |       3 |              3 | Headquarters  |
|          7 |       3 |              3 | Woburn Branch |
|          7 |       3 |              3 | Quincy Branch |
|          7 |       3 |              3 | So. NH Branch |
|          8 |       3 |              3 | Headquarters  |
```

```
|           8 |       3 |               3 | Woburn Branch  |
|           8 |       3 |               3 | Quincy Branch  |
|           8 |       3 |               3 | So. NH Branch  |
|          10 |       4 |               1 | Headquarters   |
|          10 |       4 |               1 | Woburn Branch  |
|          10 |       4 |               1 | Quincy Branch  |
|          10 |       4 |               1 | So. NH Branch  |
...
...
...
|          24 |      10 |               4 | Headquarters   |
|          24 |      10 |               4 | Woburn Branch  |
|          24 |      10 |               4 | Quincy Branch  |
|          24 |      10 |               4 | So. NH Branch  |
|          25 |      10 |               4 | Headquarters   |
|          25 |      10 |               4 | Woburn Branch  |
|          25 |      10 |               4 | Quincy Branch  |
|          25 |      10 |               4 | So. NH Branch  |
|          27 |      11 |               2 | Headquarters   |
|          27 |      11 |               2 | Woburn Branch  |
|          27 |      11 |               2 | Quincy Branch  |
|          27 |      11 |               2 | So. NH Branch  |
|          28 |      12 |               4 | Headquarters   |
|          28 |      12 |               4 | Woburn Branch  |
|          28 |      12 |               4 | Quincy Branch  |
|          28 |      12 |               4 | So. NH Branch  |
|          29 |      13 |               3 | Headquarters   |
|          29 |      13 |               3 | Woburn Branch  |
|          29 |      13 |               3 | Quincy Branch  |
|          29 |      13 |               3 | So. NH Branch  |
+-------------+---------+-----------------+----------------+
96 rows in set (0.07 sec)
```

It looks like something has gone wrong; the query should return no more than 24 rows, since there are 24 rows in the account table. What has happened is that, since the server couldn't find two identically named columns in the two tables, no join condition was generated and the two tables were cross-joined instead, resulting in 96 rows (24 accounts × 4 branches).

So, is the reduced wear and tear on the old fingers from not having to type the join condition worth the trouble? Absolutely not; you should avoid this join type and use inner joins with explicit join conditions.

Test Your Knowledge

The following exercises test your understanding of outer and cross joins. Please see Appendix C for solutions.

Exercise 10-1

Write a query that returns all product names along with the accounts based on that product (use the product_cd column in the account table to link to the product table). Include all products, even if no accounts have been opened for that product.

Exercise 10-2

Reformulate your query from Exercise 10-1 to use the other outer join type (e.g., if you used a left outer join in Exercise 10-1, use a right outer join this time) such that the results are identical to Exercise 10-1.

Exercise 10-3

Outer-join the account table to both the individual and business tables (via the account.cust_id column) such that the result set contains one row per account. Columns to include are account.account_id, account.product_cd, individual.fname, individual.lname, and business.name.

Exercise 10-4 (Extra Credit)

Devise a query that will generate the set {1, 2, 3,..., 99, 100}. (Hint: use a cross join with at least two from clause subqueries.)

CHAPTER 11

Conditional Logic

In certain situations, you may want your SQL logic to branch in one direction or another depending on the values of certain columns or expressions. This chapter focuses on how to write statements that can behave differently depending on the data encountered during statement execution.

What Is Conditional Logic?

Conditional logic is simply the ability to take one of several paths during program execution. For example, when querying customer information, you might want to retrieve either the fname/lname columns from the individual table or the name column from the business table depending on what type of customer is encountered. Using outer joins, you could return both strings and let the caller figure out which one to use, as in:

```
mysql> SELECT c.cust_id, c.fed_id, c.cust_type_cd,
    ->   CONCAT(i.fname, ' ', i.lname) indiv_name,
    ->   b.name business_name
    -> FROM customer c LEFT OUTER JOIN individual i
    ->   ON c.cust_id = i.cust_id
    ->   LEFT OUTER JOIN business b
    ->   ON c.cust_id = b.cust_id;
+---------+-------------+--------------+----------------+----------------------+
| cust_id | fed_id      | cust_type_cd | indiv_name     | business_name        |
+---------+-------------+--------------+----------------+----------------------+
|       1 | 111-11-1111 | I            | James Hadley   | NULL                 |
|       2 | 222-22-2222 | I            | Susan Tingley  | NULL                 |
|       3 | 333-33-3333 | I            | Frank Tucker   | NULL                 |
|       4 | 444-44-4444 | I            | John Hayward   | NULL                 |
|       5 | 555-55-5555 | I            | Charles Frasier| NULL                 |
|       6 | 666-66-6666 | I            | John Spencer   | NULL                 |
|       7 | 777-77-7777 | I            | Margaret Young | NULL                 |
|       8 | 888-88-8888 | I            | Louis Blake    | NULL                 |
|       9 | 999-99-9999 | I            | Richard Farley | NULL                 |
|      10 | 04-1111111  | B            | NULL           | Chilton Engineering  |
|      11 | 04-2222222  | B            | NULL           | Northeast Cooling Inc.|
|      12 | 04-3333333  | B            | NULL           | Superior Auto Body   |
```

```
|      13 | 04-4444444  | B               | NULL            | AAA Insurance Inc.    |
+---------+-------------+-----------------+-----------------+-----------------------+
13 rows in set (0.13 sec)
```

The caller can look at the value of the cust_type_cd column and decide whether to use the indiv_name or business_name column. Instead, however, you could use conditional logic via a *case expression* to determine the type of customer and return the appropriate string, as in:

```
mysql> SELECT c.cust_id, c.fed_id,
    ->   CASE
    ->     WHEN c.cust_type_cd = 'I'
    ->       THEN CONCAT(i.fname, ' ', i.lname)
    ->     WHEN c.cust_type_cd = 'B'
    ->       THEN b.name
    ->     ELSE 'Unknown'
    ->   END name
    -> FROM customer c LEFT OUTER JOIN individual i
    ->   ON c.cust_id = i.cust_id
    ->   LEFT OUTER JOIN business b
    ->   ON c.cust_id = b.cust_id;
+---------+-------------+------------------------+
| cust_id | fed_id      | name                   |
+---------+-------------+------------------------+
|       1 | 111-11-1111 | James Hadley           |
|       2 | 222-22-2222 | Susan Tingley          |
|       3 | 333-33-3333 | Frank Tucker           |
|       4 | 444-44-4444 | John Hayward           |
|       5 | 555-55-5555 | Charles Frasier        |
|       6 | 666-66-6666 | John Spencer           |
|       7 | 777-77-7777 | Margaret Young         |
|       8 | 888-88-8888 | Louis Blake            |
|       9 | 999-99-9999 | Richard Farley         |
|      10 | 04-1111111  | Chilton Engineering    |
|      11 | 04-2222222  | Northeast Cooling Inc. |
|      12 | 04-3333333  | Superior Auto Body     |
|      13 | 04-4444444  | AAA Insurance Inc.     |
+---------+-------------+------------------------+
13 rows in set (0.00 sec)
```

This version of the query returns a single name column that is generated by the *case expression starting on the second line of the query*, which, in this example, checks the value of the cust_type_cd column and returns either the individual's first/last names *or* the business name.

The Case Expression

All of the major database servers include built-in functions designed to mimic the if-then-else statement found in most programming languages (examples include Oracle's decode() function, MySQL's if() function, and SQL Server's coalesce() function). Case expressions are also designed to facilitate if-then-else logic but enjoy two advantages over built-in functions:

- The case expression is part of the SQL standard (SQL92 release) and has been implemented by Oracle Database, SQL Server, MySQL, Sybase, PostgreSQL, IBM UDB, and others.

- Case expressions are built into the SQL grammar and can be included in `select`, `insert`, `update`, and `delete` statements.

The next two subsections introduce the two different types of case expressions, and then I show you some examples of case expressions in action.

Searched Case Expressions

The case expression demonstrated earlier in the chapter is an example of a *searched case expression*, which has the following syntax:

```
CASE
    WHEN C1 THEN E1
    WHEN C2 THEN E2
    ...
    WHEN CN THEN EN
    [ELSE ED]
END
```

In the previous definition, the symbols C1, C2,..., CN represent conditions, and the symbols E1, E2,..., EN represent expressions to be returned by the case expression. If the condition in a when clause evaluates to true, then the case expression returns the corresponding expression. Additionally, the ED symbol represents the default expression, which the case expression returns if *none* of the conditions C1, C2,..., CN evaluate to true (the else clause is optional, which is why it is enclosed in square brackets). All the expressions returned by the various when clauses must evaluate to the same type (e.g., date, number, varchar).

Here's an example of a searched case expression:

```
CASE
    WHEN employee.title = 'Head Teller'
        THEN 'Head Teller'
    WHEN employee.title = 'Teller'
        AND YEAR(employee.start_date) > 2007
        THEN 'Teller Trainee'
    WHEN employee.title = 'Teller'
        AND YEAR(employee.start_date) < 2006
        THEN 'Experienced Teller'
    WHEN employee.title = 'Teller'
        THEN 'Teller'
    ELSE 'Non-Teller'
END
```

This case expression returns a string that can be used to determine hourly pay scales, print name badges, and so forth. When the case expression is evaluated, the when clauses are evaluated in order from top to bottom; as soon as one of the conditions in a when clause evaluates to true, the corresponding expression is returned and any remaining when clauses are ignored. If none of the when clause conditions evaluate to true, then the expression in the else clause is returned.

Although the previous example returns string expressions, keep in mind that case expressions may return any type of expression, including subqueries. Here's another version of the individual/business name query from earlier in the chapter that uses subqueries instead of outer joins to retrieve data from the individual and business tables:

```
mysql> SELECT c.cust_id, c.fed_id,
    ->   CASE
    ->     WHEN c.cust_type_cd = 'I' THEN
    ->       (SELECT CONCAT(i.fname, ' ', i.lname)
    ->        FROM individual i
    ->        WHERE i.cust_id = c.cust_id)
    ->     WHEN c.cust_type_cd = 'B' THEN
    ->       (SELECT b.name
    ->        FROM business b
    ->        WHERE b.cust_id = c.cust_id)
    ->     ELSE 'Unknown'
    ->   END name
    -> FROM customer c;
+---------+-------------+------------------------+
| cust_id | fed_id      | name                   |
+---------+-------------+------------------------+
|       1 | 111-11-1111 | James Hadley           |
|       2 | 222-22-2222 | Susan Tingley          |
|       3 | 333-33-3333 | Frank Tucker           |
|       4 | 444-44-4444 | John Hayward           |
|       5 | 555-55-5555 | Charles Frasier        |
|       6 | 666-66-6666 | John Spencer           |
|       7 | 777-77-7777 | Margaret Young         |
|       8 | 888-88-8888 | Louis Blake            |
|       9 | 999-99-9999 | Richard Farley         |
|      10 | 04-1111111  | Chilton Engineering    |
|      11 | 04-2222222  | Northeast Cooling Inc. |
|      12 | 04-3333333  | Superior Auto Body     |
|      13 | 04-4444444  | AAA Insurance Inc.     |
+---------+-------------+------------------------+
13 rows in set (0.01 sec)
```

This version of the query includes only the customer table in the from clause and uses correlated subqueries to retrieve the appropriate name for each customer. I prefer this version over the outer join version from earlier in the chapter, since the server reads from the individual and business tables only as needed instead of always joining all three tables.

Simple Case Expressions

The *simple case expression* is quite similar to the searched case expression but is a bit less flexible. Here's the syntax:

```
CASE V0
  WHEN V1 THEN E1
  WHEN V2 THEN E2
  ...
  WHEN VN THEN EN
  [ELSE ED]
END
```

In the preceding definition, V0 represents a value, and the symbols V1, V2,..., VN represent values that are to be compared to V0. The symbols E1, E2,..., EN represent expressions to be returned by the case expression, and ED represents the expression to be returned if none of the values in the set V1, V2,..., VN match the V0 value.

Here's an example of a simple case expression:

```
CASE customer.cust_type_cd
  WHEN 'I' THEN
    (SELECT CONCAT(i.fname, ' ', i.lname)
     FROM individual I
     WHERE i.cust_id = customer.cust_id)
  WHEN 'B' THEN
    (SELECT b.name
     FROM business b
     WHERE b.cust_id = customer.cust_id)
  ELSE 'Unknown Customer Type'
END
```

Simple case expressions are less powerful than searched case expressions because you can't specify your own conditions; instead, equality conditions are built for you. To show you what I mean, here's a searched case expression having the same logic as the previous simple case expression:

```
CASE
  WHEN customer.cust_type_cd = 'I' THEN
    (SELECT CONCAT(i.fname, ' ', i.lname)
     FROM individual I
     WHERE i.cust_id = customer.cust_id)
  WHEN customer.cust_type_cd = 'B' THEN
    (SELECT b.name
     FROM business b
     WHERE b.cust_id = customer.cust_id)
  ELSE 'Unknown Customer Type'
END
```

With searched case expressions, you can build range conditions, inequality conditions, and multipart conditions using and/or/not, so I would recommend using searched case expressions for all but the simplest logic.

Case Expression Examples

The following sections present a variety of examples illustrating the utility of conditional logic in SQL statements.

Result Set Transformations

You may have run into a situation where you are performing aggregations over a finite set of values, such as days of the week, but you want the result set to contain a single row with one column per value instead of one row per value. As an example, let's say you have been asked to write a query that shows the number of accounts opened in the years 2000 through 2005:

```
mysql> SELECT YEAR(open_date) year, COUNT(*) how_many
    -> FROM account
    -> WHERE open_date > '1999-12-31'
    ->   AND open_date < '2006-01-01'
    -> GROUP BY YEAR(open_date);
+------+----------+
| year | how_many |
+------+----------+
| 2000 |        3 |
| 2001 |        4 |
| 2002 |        5 |
| 2003 |        3 |
| 2004 |        9 |
+------+----------+
5 rows in set (0.00 sec)
```

However, you have also been instructed to return a single row of data with six columns (one for each year in the data range). To transform this result set into a single row, you will need to create six columns and, within each column, sum *only* those rows pertaining to the year in question:

```
mysql> SELECT
    ->     SUM(CASE
    ->         WHEN EXTRACT(YEAR FROM open_date) = 2000 THEN 1
    ->         ELSE 0
    ->         END) year_2000,
    ->     SUM(CASE
    ->         WHEN EXTRACT(YEAR FROM open_date) = 2001 THEN 1
    ->         ELSE 0
    ->         END) year_2001,
    ->     SUM(CASE
    ->         WHEN EXTRACT(YEAR FROM open_date) = 2002 THEN 1
    ->         ELSE 0
    ->         END) year_2002,
    ->     SUM(CASE
    ->         WHEN EXTRACT(YEAR FROM open_date) = 2003 THEN 1
    ->         ELSE 0
    ->         END) year_2003,
    ->     SUM(CASE
    ->         WHEN EXTRACT(YEAR FROM open_date) = 2004 THEN 1
    ->         ELSE 0
    ->         END) year_2004,
    ->     SUM(CASE
    ->         WHEN EXTRACT(YEAR FROM open_date) = 2005 THEN 1
    ->         ELSE 0
    ->         END) year_2005
```

```
    -> FROM account
    -> WHERE open_date > '1999-12-31' AND open_date < '2006-01-01';
+-----------+-----------+-----------+-----------+-----------+-----------+
| year_2000 | year_2001 | year_2002 | year_2003 | year_2004 | year_2005 |
+-----------+-----------+-----------+-----------+-----------+-----------+
|         3 |         4 |         5 |         3 |         9 |         0 |
+-----------+-----------+-----------+-----------+-----------+-----------+
1 row in set (0.01 sec)
```

Each of the six columns in the previous query are identical, except for the year value. When the `extract()` function returns the desired year for that column, the case expression returns the value 1; otherwise, it returns a 0. When summed over all accounts opened since 2000, each column returns the number of accounts opened for that year. Obviously, such transformations are practical for only a small number of values; generating one column for each year since 1905 would quickly become tedious.

 Although it is a bit advanced for this book, it is worth pointing out that both SQL Server and Oracle Database 11g include PIVOT clauses specifically for these types of queries.

Selective Aggregation

Back in Chapter 9, I showed a partial solution for an example that demonstrated how to find accounts whose account balances don't agree with the raw data in the `transaction` table. The reason for the partial solution was that a full solution requires the use of conditional logic, so all the pieces are now in place to finish the job. Here's where I left off in Chapter 9:

```
SELECT CONCAT('ALERT! : Account #', a.account_id,
  ' Has Incorrect Balance!')
FROM account a
WHERE (a.avail_balance, a.pending_balance) <>
 (SELECT SUM(<expression to generate available balance>),
    SUM(<expression to generate pending balance>)
  FROM transaction t
  WHERE t.account_id = a.account_id);
```

The query uses a correlated subquery on the `transaction` table to sum together the individual transactions for a given account. When summing transactions, you need to consider the following two issues:

- Transaction amounts are always positive, so you need to look at the transaction type to see whether the transaction is a debit or a credit and flip the sign (multiply by –1) for debit transactions.
- If the date in the `funds_avail_date` column is greater than the current day, the transaction should be added to the pending balance total but not to the available balance total.

While some transactions need to be excluded from the available balance, all transactions are included in the pending balance, making it the simpler of the two calculations. Here's the case expression used to calculate the pending balance:

```
CASE
  WHEN transaction.txn_type_cd = 'DBT'
    THEN transaction.amount * -1
  ELSE transaction.amount
END
```

Thus, all transaction amounts are multiplied by –1 for debit transactions and are left as is for credit transactions. This same logic applies to the available balance calculation as well, but only transactions that have become available should be included. Therefore, the case expression used to calculate available balance includes one additional when clause:

```
CASE
  WHEN transaction.funds_avail_date > CURRENT_TIMESTAMP()
    THEN 0
  WHEN transaction.txn_type_cd = 'DBT'
    THEN transaction.amount * -1
  ELSE transaction.amount
END
```

With the first when clause in place, unavailable funds, such as checks that have not cleared, will contribute $0 to the sum. Here's the final query with the two case expressions in place:

```
SELECT CONCAT('ALERT! : Account #', a.account_id,
  ' Has Incorrect Balance!')
FROM account a
WHERE (a.avail_balance, a.pending_balance) <>
 (SELECT
    SUM(CASE
          WHEN t.funds_avail_date > CURRENT_TIMESTAMP()
            THEN 0
          WHEN t.txn_type_cd = 'DBT'
            THEN t.amount * -1
          ELSE t.amount
        END),
    SUM(CASE
          WHEN t.txn_type_cd = 'DBT'
            THEN t.amount * -1
          ELSE t.amount
        END)
  FROM transaction t
  WHERE t.account_id = a.account_id);
```

By using conditional logic, the sum() aggregate functions are being fed manipulated data by the two case expressions, allowing the appropriate amounts to be summed.

Checking for Existence

Sometimes you will want to determine whether a relationship exists between two entities without regard for the quantity. For example, you might want to know whether a customer has any checking or savings accounts, but you don't care whether a customer has more than one of each type of account. Here's a query that uses multiple case expressions to generate two output columns, one to show whether the customer has any checking accounts and the other to show whether the customer has any savings accounts:

```
mysql> SELECT c.cust_id, c.fed_id, c.cust_type_cd,
    ->   CASE
    ->     WHEN EXISTS (SELECT 1 FROM account a
    ->       WHERE a.cust_id = c.cust_id
    ->         AND a.product_cd = 'CHK') THEN 'Y'
    ->     ELSE 'N'
    ->   END has_checking,
    ->   CASE
    ->     WHEN EXISTS (SELECT 1 FROM account a
    ->       WHERE a.cust_id = c.cust_id
    ->         AND a.product_cd = 'SAV') THEN 'Y'
    ->     ELSE 'N'
    ->   END has_savings
    -> FROM customer c;
```

cust_id	fed_id	cust_type_cd	has_checking	has_savings
1	111-11-1111	I	Y	Y
2	222-22-2222	I	Y	Y
3	333-33-3333	I	Y	N
4	444-44-4444	I	Y	Y
5	555-55-5555	I	Y	N
6	666-66-6666	I	Y	N
7	777-77-7777	I	N	N
8	888-88-8888	I	Y	Y
9	999-99-9999	I	Y	N
10	04-1111111	B	Y	N
11	04-2222222	B	N	N
12	04-3333333	B	Y	N
13	04-4444444	B	N	N

13 rows in set (0.00 sec)

Each case expression includes a correlated subquery against the account table; one looks for checking accounts, the other for savings accounts. Since each when clause uses the exists operator, the conditions evaluate to true as long as the customer has at least one of the desired accounts.

In other cases, you may care how many rows are encountered, but only up to a point. For example, the next query uses a simple case expression to count the number of accounts for each customer, and then returns either 'None', '1', '2', or '3+':

```
mysql> SELECT c.cust_id, c.fed_id, c.cust_type_cd,
    ->    CASE (SELECT COUNT(*) FROM account a
    ->        WHERE a.cust_id = c.cust_id)
    ->      WHEN 0 THEN 'None'
    ->      WHEN 1 THEN '1'
    ->      WHEN 2 THEN '2'
    ->      ELSE '3+'
    ->    END num_accounts
    -> FROM customer c;
+---------+-------------+--------------+--------------+
| cust_id | fed_id      | cust_type_cd | num_accounts |
+---------+-------------+--------------+--------------+
|       1 | 111-11-1111 | I            | 3+           |
|       2 | 222-22-2222 | I            | 2            |
|       3 | 333-33-3333 | I            | 2            |
|       4 | 444-44-4444 | I            | 3+           |
|       5 | 555-55-5555 | I            | 1            |
|       6 | 666-66-6666 | I            | 2            |
|       7 | 777-77-7777 | I            | 1            |
|       8 | 888-88-8888 | I            | 2            |
|       9 | 999-99-9999 | I            | 3+           |
|      10 | 04-1111111  | B            | 2            |
|      11 | 04-2222222  | B            | 1            |
|      12 | 04-3333333  | B            | 1            |
|      13 | 04-4444444  | B            | 1            |
+---------+-------------+--------------+--------------+
13 rows in set (0.01 sec)
```

For this query, I didn't want to differentiate between customers having more than two accounts, so the case expression simply creates a '3+' category. Such a query might be useful if you were looking for customers to contact regarding opening a new account with the bank.

Division-by-Zero Errors

When performing calculations that include division, you should always take care to ensure that the denominators are never equal to zero. Whereas some database servers, such as Oracle Database, will throw an error when a zero denominator is encountered, MySQL simply sets the result of the calculation to null, as demonstrated by the following:

```
mysql> SELECT 100 / 0;
+---------+
| 100 / 0 |
+---------+
|    NULL |
+---------+
1 row in set (0.00 sec)
```

To safeguard your calculations from encountering errors or, even worse, from being mysteriously set to null, you should wrap all denominators in conditional logic, as demonstrated by the following:

```
mysql> SELECT a.cust_id, a.product_cd, a.avail_balance /
    ->    CASE
    ->      WHEN prod_tots.tot_balance = 0 THEN 1
    ->      ELSE prod_tots.tot_balance
    ->    END percent_of_total
    -> FROM account a INNER JOIN
    ->   (SELECT a.product_cd, SUM(a.avail_balance) tot_balance
    ->    FROM account a
    ->    GROUP BY a.product_cd) prod_tots
    ->    ON a.product_cd = prod_tots.product_cd;
+---------+------------+-------------------+
| cust_id | product_cd | percent_of_total  |
+---------+------------+-------------------+
|      10 | BUS        |          0.000000 |
|      11 | BUS        |          1.000000 |
|       1 | CD         |          0.153846 |
|       6 | CD         |          0.512821 |
|       7 | CD         |          0.256410 |
|       9 | CD         |          0.076923 |
|       1 | CHK        |          0.014488 |
|       2 | CHK        |          0.030928 |
|       3 | CHK        |          0.014488 |
|       4 | CHK        |          0.007316 |
|       5 | CHK        |          0.030654 |
|       6 | CHK        |          0.001676 |
|       8 | CHK        |          0.047764 |
|       9 | CHK        |          0.001721 |
|      10 | CHK        |          0.322911 |
|      12 | CHK        |          0.528052 |
|       3 | MM         |          0.129802 |
|       4 | MM         |          0.321915 |
|       9 | MM         |          0.548282 |
|       1 | SAV        |          0.269431 |
|       2 | SAV        |          0.107773 |
|       4 | SAV        |          0.413723 |
|       8 | SAV        |          0.209073 |
|      13 | SBL        |          1.000000 |
+---------+------------+-------------------+
24 rows in set (0.13 sec)
```

This query computes the ratio of each account balance to the total balance for all accounts of the same product type. Since some product types, such as business loans,
could have a total balance of zero if all loans were currently paid in full, it is best to
include the case expression to ensure that the denominator is never zero.

Conditional Updates

When updating rows in a table, you sometimes need to decide what values to set certain
columns to. For example, after inserting a new transaction, you need to modify the
avail_balance, pending_balance, and last_activity_date columns in the account table.
Although the last two columns are easily updated, to correctly modify the
avail_balance column you need to know whether the funds from the transaction are

immediately available by checking the funds_avail_date column in the transaction table. Given that transaction ID 999 has just been inserted, you can use the following update statement to modify the three columns in the account table:

```
1   UPDATE account
2     SET last_activity_date = CURRENT_TIMESTAMP(),
3       pending_balance = pending_balance +
4       (SELECT t.amount *
5          CASE t.txn_type_cd WHEN 'DBT' THEN -1 ELSE 1 END
6        FROM transaction t
7        WHERE t.txn_id = 999),
8       avail_balance = avail_balance +
9       (SELECT
10          CASE
11            WHEN t.funds_avail_date > CURRENT_TIMESTAMP() THEN 0
12            ELSE t.amount *
13              CASE t.txn_type_cd WHEN 'DBT' THEN -1 ELSE 1 END
14          END
15        FROM transaction t
16        WHERE t.txn_id = 999)
17   WHERE account.account_id =
18     (SELECT t.account_id
19      FROM transaction t
20      WHERE t.txn_id = 999);
```

This statement contains a total of three case expressions : two of them (lines 5 and 13) are used to flip the sign on the transaction amount for debit transactions, and the third case expression (line 10) is used to check the funds availability date. If the date is in the future, then zero is added to the available balance; otherwise, the transaction amount is added.

Handling Null Values

While null values are the appropriate thing to store in a table if the value for a column is unknown, it is not always appropriate to retrieve null values for display or to take part in expressions. For example, you might want to display the word *unknown* on a data entry screen rather than leaving a field blank. When retrieving the data, you can use a case expression to substitute the string if the value is null, as in:

```
SELECT emp_id, fname, lname,
  CASE
    WHEN title IS NULL THEN 'Unknown'
    ELSE title
  END
FROM employee;
```

For calculations, null values often cause a null result, as demonstrated by the following:

```
mysql> SELECT (7 * 5) / ((3 + 14) * null);
+-----------------------------+
| (7 * 5) / ((3 + 14) * null) |
+-----------------------------+
|                        NULL |
```

```
+-----------------------------+
```
```
1 row in set (0.08 sec)
```

When performing calculations, case expressions are useful for translating a `null` value into a number (usually 0 or 1) that will allow the calculation to yield a non-`null` value. If you are performing a calculation that includes the `account.avail_balance` column, for example, you could substitute a 0 (if doing addition or subtraction) or a 1 (if doing multiplication or division) for those accounts that have been established but haven't yet been funded:

```
SELECT <some calculation> +
  CASE
    WHEN avail_balance IS NULL THEN 0
    ELSE avail_balance
  END
  + <rest of calculation>
...
```

If a numeric column is allowed to contain `null` values, it is generally a good idea to use conditional logic in any calculations that include the column so that the results are usable.

Test Your Knowledge

Challenge your ability to work through conditional logic problems with the examples that follow. When you're done, compare your solutions with those in Appendix C.

Exercise 11-1

Rewrite the following query, which uses a simple case expression, so that the same results are achieved using a searched case expression. Try to use as few when clauses as possible.

```
SELECT emp_id,
  CASE title
    WHEN 'President' THEN 'Management'
    WHEN 'Vice President' THEN 'Management'
    WHEN 'Treasurer' THEN 'Management'
    WHEN 'Loan Manager' THEN 'Management'
    WHEN 'Operations Manager' THEN 'Operations'
    WHEN 'Head Teller' THEN 'Operations'
    WHEN 'Teller' THEN 'Operations'
    ELSE 'Unknown'
  END
FROM employee;
```

Exercise 11-2

Rewrite the following query so that the result set contains a single row with four columns (one for each branch). Name the four columns branch_1 through branch_4.

```
mysql> SELECT open_branch_id, COUNT(*)
    -> FROM account
    -> GROUP BY open_branch_id;
+----------------+----------+
| open_branch_id | COUNT(*) |
+----------------+----------+
|              1 |        8 |
|              2 |        7 |
|              3 |        3 |
|              4 |        6 |
+----------------+----------+
4 rows in set (0.00 sec)
```

CHAPTER 12

Transactions

All of the examples thus far in this book have been individual, independent SQL statements. While this may be the norm for ad hoc reporting or data maintenance scripts, application logic will frequently include multiple SQL statements that need to execute together as a logical unit of work. This chapter explores the need and the infrastructure necessary to execute multiple SQL statements concurrently.

Multiuser Databases

Database management systems allow not only a single user to query and modify data, but multiple people to do so simultaneously. If every user is only executing queries, such as might be the case with a data warehouse during normal business hours, then there are very few issues for the database server to deal with. If some of the users are adding and/or modifying data, however, the server must handle quite a bit more bookkeeping.

Let's say, for example, that you are running a report that shows the available balance for all the checking accounts opened at your branch. At the same time you are running the report, however, the following activities are occurring:

- A teller at your branch is handling a deposit for one of your customers.
- A customer is finishing a withdrawal at the ATM in the front lobby.
- The bank's month-end application is applying interest to the accounts.

While your report is running, therefore, multiple users are modifying the underlying data, so what numbers should appear on the report? The answer depends somewhat on how your server handles *locking*, which is described in the next section.

Locking

Locks are the mechanism the database server uses to control simultaneous use of data resources. When some portion of the database is locked, any other users wishing to modify (or possibly read) that data must wait until the lock has been released. Most database servers use one of two locking strategies:

- Database writers must request and receive from the server a *write lock* to modify data, and database readers must request and receive from the server a *read lock* to query data. While multiple users can read data simultaneously, only one write lock is given out at a time for each table (or portion thereof), and read requests are blocked until the write lock is released.

- Database writers must request and receive from the server a write lock to modify data, but readers do not need any type of lock to query data. Instead, the server ensures that a reader sees a consistent view of the data (the data seems the same even though other users may be making modifications) from the time her query begins until her query has finished. This approach is known as *versioning*.

There are pros and cons to both approaches. The first approach can lead to long wait times if there are many concurrent read and write requests, and the second approach can be problematic if there are long-running queries while data is being modified. Of the three servers discussed in this book, Microsoft SQL Server uses the first approach, Oracle Database uses the second approach, and MySQL uses both approaches (depending on your choice of *storage engine*, which we'll discuss a bit later in the chapter).

Lock Granularities

There are also a number of different strategies that you may employ when deciding *how* to lock a resource. The server may apply a lock at one of three different levels, or *granularities*:

Table locks
   Keep multiple users from modifying data in the same table simultaneously

Page locks
   Keep multiple users from modifying data on the same page (a page is a segment of memory generally in the range of 2 KB to 16 KB) of a table simultaneously

Row locks
   Keep multiple users from modifying the same row in a table simultaneously

Again, there are pros and cons to these approaches. It takes very little bookkeeping to lock entire tables, but this approach quickly yields unacceptable wait times as the number of users increases. On the other hand, row locking takes quite a bit more bookkeeping, but it allows many users to modify the same table as long as they are interested in different rows. Of the three servers discussed in this book, Microsoft SQL Server uses page, row, and table locking, Oracle Database uses only row locking, and MySQL uses table, page, or row locking (depending, again, on your choice of storage engine). SQL Server will, under certain circumstances, *escalate* locks from row to page, and from page to table, whereas Oracle Database will never escalate locks.

To get back to your report, the data that appears on the pages of the report will mirror either the state of the database when your report started (if your server uses a versioning approach) or the state of the database when the server issues the reporting application a read lock (if your server uses both read and write locks).

What Is a Transaction?

If database servers enjoyed 100% uptime, if users always allowed programs to finish executing, and if applications always completed without encountering fatal errors that halt execution, then there would be nothing left to discuss regarding concurrent database access. However, we can rely on none of these things, so one more element is necessary to allow multiple users to access the same data.

This extra piece of the concurrency puzzle is the *transaction*, which is a device for grouping together multiple SQL statements such that either *all* or *none* of the statements succeed (a property known as *atomicity*). If you attempt to transfer $500 from your savings account to your checking account, you would be a bit upset if the money were successfully withdrawn from your savings account but never made it to your checking account. Whatever the reason for the failure (the server was shut down for maintenance, the request for a page lock on the account table timed out, etc.), you want your $500 back.

To protect against this kind of error, the program that handles your transfer request would first begin a transaction, then issue the SQL statements needed to move the money from your savings to your checking account, and, if everything succeeds, end the transaction by issuing the commit command. If something unexpected happens, however, the program would issue a rollback command, which instructs the server to undo all changes made since the transaction began. The entire process might look something like the following:

```
START TRANSACTION;

 /* withdraw money from first account, making sure balance is sufficient */
UPDATE account SET avail_balance = avail_balance - 500
WHERE account_id = 9988
  AND avail_balance > 500;

IF <exactly one row was updated by the previous statement> THEN
  /* deposit money into second account */
  UPDATE account SET avail_balance = avail_balance + 500
    WHERE account_id = 9989;

  IF <exactly one row was updated by the previous statement> THEN
    /* everything worked, make the changes permanent */
    COMMIT;
  ELSE
    /* something went wrong, undo all changes in this transaction */
    ROLLBACK;
  END IF;
```

```
ELSE
  /* insufficient funds, or error encountered during update */
  ROLLBACK;
END IF;
```

 While the previous code block may look similar to one of the procedural languages provided by the major database companies, such as Oracle's PL/SQL or Microsoft's Transact-SQL, it is written in pseudocode and does not attempt to mimic any particular language.

The previous code block begins by starting a transaction and then attempts to remove $500 from the checking account and add it to the savings account. If all goes well, the transaction is committed; if anything goes awry, however, the transaction is rolled back, meaning that all data changes since the beginning of the transaction are undone.

By using a transaction, the program ensures that your $500 either stays in your savings account or moves to your checking account, without the possibility of it falling into a crack. Regardless of whether the transaction was committed or was rolled back, all resources acquired (e.g., write locks) during the execution of the transaction are released when the transaction completes.

Of course, if the program manages to complete both `update` statements but the server shuts down before a `commit` or `rollback` can be executed, then the transaction will be rolled back when the server comes back online. (One of the tasks that a database server must complete before coming online is to find any incomplete transactions that were underway when the server shut down and to roll them back.) Additionally, if your program finishes a transaction and issues a `commit`, but the server shuts down before the changes have been applied to permanent storage (i.e., the modified data is sitting in memory but has not been flushed to disk), then the database server must reapply the changes from your transaction when the server is restarted (a property known as *durability*).

Starting a Transaction

Database servers handle transaction creation in one of two ways:

- An active transaction is always associated with a database session, so there is no need or method to explicitly begin a transaction. When the current transaction ends, the server automatically begins a new transaction for your session.
- Unless you explicitly begin a transaction, individual SQL statements are automatically committed independently of one another. To begin a transaction, you must first issue a command.

Of the three servers, Oracle Database takes the first approach, while Microsoft SQL Server and MySQL take the second approach. One of the advantages of Oracle's approach to transactions is that, even if you are issuing only a single SQL command, you have the ability to roll back the changes if you don't like the outcome or if you change your mind. Thus, if you forget to add a `where` clause to your `delete` statement, you will have the opportunity to undo the damage (assuming you've had your morning coffee and realize that you didn't mean to delete all 125,000 rows in your table). With MySQL and SQL Server, however, once you press the Enter key, the changes brought about by your SQL statement will be permanent (unless your DBA can retrieve the original data from a backup or from some other means).

The SQL:2003 standard includes a `start transaction` command to be used when you want to explicitly begin a transaction. While MySQL conforms to the standard, SQL Server users must instead issue the command `begin transaction`. With both servers, until you explicitly begin a transaction, you are in what is known as *auto-commit mode*, which means that individual statements are automatically committed by the server. You can, therefore, decide that you want to be in a transaction and issue a start/begin transaction command, or you can simply let the server commit individual statements.

Both MySQL and SQL Server allow you to turn off auto-commit mode for individual sessions, in which case, the servers will act just like Oracle Database regarding transactions. With SQL Server, you issue the following command to disable auto-commit mode:

```
SET IMPLICIT_TRANSACTIONS ON
```

MySQL allows you to disable auto-commit mode via the following:

```
SET AUTOCOMMIT=0
```

Once you have left auto-commit mode, all SQL commands take place within the scope of a transaction and must be explicitly committed or rolled back.

 A word of advice: shut off auto-commit mode each time you log in, and get in the habit of running all of your SQL statements within a transaction. If nothing else, it may save you the embarrassment of having to ask your DBA to reconstruct data that you have inadvertently deleted.

Ending a Transaction

Once a transaction has begun, whether explicitly via the `start transaction` command or implicitly by the database server, you must explicitly end your transaction for your changes to become permanent. You do this by way of the `commit` command, which instructs the server to mark the changes as permanent and release any resources (i.e., page or row locks) used during the transaction.

If you decide that you want to undo all the changes made since starting the transaction, you must issue the `rollback` command, which instructs the server to return the data to its pretransaction state. After the `rollback` has been completed, any resources used by your session are released.

Along with issuing either the commit or rollback command, there are several other scenarios by which your transaction can end, either as an indirect result of your actions or as a result of something outside your control:

- The server shuts down, in which case, your transaction will be rolled back automatically when the server is restarted.
- You issue an SQL schema statement, such as alter table, which will cause the current transaction to be committed and a new transaction to be started.
- You issue another start transaction command, which will cause the previous transaction to be committed.
- The server prematurely ends your transaction because the server detects a *deadlock* and decides that your transaction is the culprit. In this case, the transaction will be rolled back and you will receive an error message.

Of these four scenarios, the first and third are fairly straightforward, but the other two merit some discussion. As far as the second scenario is concerned, alterations to a database, whether it be the addition of a new table or index or the removal of a column from a table, cannot be rolled back, so commands that alter your schema must take place outside a transaction. If a transaction is currently underway, therefore, the server will commit your current transaction, execute the SQL schema statement command(s), and then automatically start a new transaction for your session. The server will not inform you of what has happened, so you should be careful that the statements that comprise a unit of work are not inadvertently broken up into multiple transactions by the server.

The fourth scenario deals with deadlock detection. A deadlock occurs when two different transactions are waiting for resources that the other transaction currently holds. For example, transaction A might have just updated the account table and is waiting for a write lock on the transaction table, while transaction B has inserted a row into the transaction table and is waiting for a write lock on the account table. If both transactions happen to be modifying the same page or row (depending on the lock granularity in use by the database server), then they will each wait forever for the other transaction to finish and free up the needed resource. Database servers must always be on the lookout for these situations so that throughput doesn't grind to a halt; when a deadlock is detected, one of the transactions is chosen (either arbitrarily or by some criteria) to be rolled back so that the other transaction may proceed. Most of the time, the terminated transaction can be restarted and will succeed without encountering another deadlock situation.

Unlike the second scenario discussed earlier, the database server will raise an error to inform you that your transaction has been rolled back due to deadlock detection. With MySQL, for example, you will receive error #1213, which carries the following message:

```
Message: Deadlock found when trying to get lock; try restarting transaction
```

As the error message suggests, it is a reasonable practice to retry a transaction that has been rolled back due to deadlock detection. However, if deadlocks become fairly common, then you may need to modify the applications that access the database to decrease the probability of deadlocks (one common strategy is to ensure that data resources are always accessed in the same order, such as always modifying account data before inserting transaction data).

Transaction Savepoints

In some cases, you may encounter an issue within a transaction that requires a rollback, but you may not want to undo *all* of the work that has transpired. For these situations, you can establish one or more *savepoints* within a transaction and use them to roll back to a particular location within your transaction rather than rolling all the way back to the start of the transaction.

Choosing a Storage Engine

When using Oracle Database or Microsoft SQL Server, a single set of code is responsible for low-level database operations, such as retrieving a particular row from a table based on primary key value. The MySQL server, however, has been designed so that multiple storage engines may be utilized to provide low-level database functionality, including resource locking and transaction management. As of version 6.0, MySQL includes the following storage engines:

MyISAM
> A nontransactional engine employing table locking

MEMORY
> A nontransactional engine used for in-memory tables

BDB
> A transactional engine employing page-level locking

InnoDB
> A transactional engine employing row-level locking

Merge
> A specialty engine used to make multiple identical MyISAM tables appear as a single table (a.k.a. table partitioning)

Maria
> A MyISAM replacement included in version 6.0.6 that adds full recovery capabilities

Falcon
> A new (as of 6.0.4) high-performance transactional engine employing row-level locking

Archive
> A specialty engine used to store large amounts of unindexed data, mainly for archival purposes

---

Although you might think that you would be forced to choose a single storage engine for your database, MySQL is flexible enough to allow you to choose a storage engine on a table-by-table basis. For any tables that might take part in transactions, however, you should choose the InnoDB or Falcon storage engine, which uses row-level locking and versioning to provide the highest level of concurrency across the different storage engines.

You may explicitly specify a storage engine when creating a table, or you can change an existing table to use a different engine. If you do not know what engine is assigned to a table, you can use the show table command, as demonstrated by the following:

```
mysql> SHOW TABLE STATUS LIKE 'transaction' \G
*************************** 1. row ***************************
           Name: transaction
         Engine: InnoDB
        Version: 10
     Row_format: Compact
           Rows: 21
 Avg_row_length: 780
    Data_length: 16384
Max_data_length: 0
   Index_length: 49152
      Data_free: 0
 Auto_increment: 22
    Create_time: 2008-02-19 23:24:36
    Update_time: NULL
     Check_time: NULL
      Collation: latin1_swedish_ci
       Checksum: NULL
 Create_options:
        Comment:
1 row in set (1.46 sec)
```

Looking at the second item, you can see that the transaction table is already using the InnoDB engine. If it were not, you could assign the InnoDB engine to the transaction table via the following command:

```
ALTER TABLE transaction ENGINE = INNODB;
```

All savepoints must be given a name, which allows you to have multiple savepoints within a single transaction. To create a savepoint named my_savepoint, you can do the following:

```
SAVEPOINT my_savepoint;
```

To roll back to a particular savepoint, you simply issue the rollback command followed by the keywords to savepoint and the name of the savepoint, as in:

```
ROLLBACK TO SAVEPOINT my_savepoint;
```

Here's an example of how savepoints may be used:

```
START TRANSACTION;

UPDATE product
SET date_retired = CURRENT_TIMESTAMP()
```

```
WHERE product_cd = 'XYZ';

SAVEPOINT before_close_accounts;

UPDATE account
SET status = 'CLOSED', close_date = CURRENT_TIMESTAMP(),
  last_activity_date = CURRENT_TIMESTAMP()
WHERE product_cd = 'XYZ';

ROLLBACK TO SAVEPOINT before_close_accounts;
COMMIT;
```

The net effect of this transaction is that the mythical XYZ product is retired but none of the accounts are closed.

When using savepoints, remember the following:

- Despite the name, nothing is saved when you create a savepoint. You must eventually issue a commit if you want your transaction to be made permanent.

- If you issue a rollback without naming a savepoint, all savepoints within the transaction will be ignored and the entire transaction will be undone.

If you are using SQL Server, you will need to use the proprietary command save transaction to create a savepoint and rollback transaction to roll back to a savepoint, with each command being followed by the savepoint name.

Test Your Knowledge

Test your understanding of transactions by working through the following exercise. When you're done, compare your solution with that in Appendix C.

Exercise 12-1

Generate a transaction to transfer $50 from Frank Tucker's money market account to his checking account. You will need to insert two rows into the transaction table and update two rows in the account table.

CHAPTER 13

Indexes and Constraints

Because the focus of this book is on programming techniques, the first 12 chapters concentrated on elements of the SQL language that you can use to craft powerful `select`, `insert`, `update`, and `delete` statements. However, other database features *indirectly* affect the code you write. This chapter focuses on two of those features: indexes and constraints.

Indexes

When you insert a row into a table, the database server does not attempt to put the data in any particular location within the table. For example, if you add a row to the `department` table, the server doesn't place the row in numeric order via the `dept_id` column or in alphabetical order via the `name` column. Instead, the server simply places the data in the next available location within the file (the server maintains a list of free space for each table). When you query the `department` table, therefore, the server will need to inspect every row of the table to answer the query. For example, let's say that you issue the following query:

```
mysql> SELECT dept_id, name
    -> FROM department
    -> WHERE name LIKE 'A%';
+---------+----------------+
| dept_id | name           |
+---------+----------------+
|       3 | Administration |
+---------+----------------+
1 row in set (0.03 sec)
```

To find all departments whose name begins with *A*, the server must visit each row in the `department` table and inspect the contents of the `name` column; if the department name begins with *A*, then the row is added to the result set. This type of access is known as a *table scan*.

While this method works fine for a table with only three rows, imagine how long it might take to answer the query if the table contains 3 million rows. At some number of rows larger than three and smaller than 3 million, a line is crossed where the server cannot answer the query within a reasonable amount of time without additional help. This help comes in the form of one or more *indexes* on the department table.

Even if you have never heard of a database index, you are certainly aware of what an index is (e.g., this book has one). An index is simply a mechanism for finding a specific item within a resource. Each technical publication, for example, includes an index at the end that allows you to locate a specific word or phrase within the publication. The index lists these words and phrases in alphabetical order, allowing the reader to move quickly to a particular letter within the index, find the desired entry, and then find the page or pages on which the word or phrase may be found.

In the same way that a person uses an index to find words within a publication, a database server uses indexes to locate rows in a table. Indexes are special tables that, unlike normal data tables, *are* kept in a specific order. Instead of containing *all* of the data about an entity, however, an index contains only the column (or columns) used to locate rows in the data table, along with information describing where the rows are physically located. Therefore, the role of indexes is to facilitate the retrieval of a subset of a table's rows and columns *without* the need to inspect every row in the table.

Index Creation

Returning to the department table, you might decide to add an index on the name column to speed up any queries that specify a full or partial department name, as well as any update or delete operations that specify a department name. Here's how you can add such an index to a MySQL database:

```
mysql> ALTER TABLE department
    -> ADD INDEX dept_name_idx (name);
Query OK, 3 rows affected (0.08 sec)
Records: 3  Duplicates: 0  Warnings: 0
```

This statement creates an index (a B-tree index to be precise, but more on this shortly) on the department.name column; furthermore, the index is given the name dept_name_idx. With the index in place, the query optimizer (which we discussed in Chapter 3) can choose to use the index if it is deemed beneficial to do so (with only three rows in the department table, for example, the optimizer might very well choose to ignore the index and read the entire table). If there is more than one index on a table, the optimizer must decide which index will be the most beneficial for a particular SQL statement.

MySQL treats indexes as optional components of a table, which is why you must use the `alter table` command to add or remove an index. Other database servers, including SQL Server and Oracle Database, treat indexes as independent schema objects. For both SQL Server and Oracle, therefore, you would generate an index using the `create index` command, as in:

```
CREATE INDEX dept_name_idx
ON department (name);
```

As of MySQL version 5.0, a `create index` command is available, although it is mapped to the `alter table` command.

All database servers allow you to look at the available indexes. MySQL users can use the `show` command to see all of the indexes on a specific table, as in:

```
mysql> SHOW INDEX FROM department \G *************************** 1. row
*************************** 1. row ***************************
        Table: department
   Non_unique: 0
     Key_name: PRIMARY
 Seq_in_index: 1
  Column_name: dept_id
    Collation: A
  Cardinality: 3
     Sub_part: NULL
       Packed: NULL
         Null:
   Index_type: BTREE
      Comment:
Index_Comment:
*************************** 2. row ***************************
        Table: department
   Non_unique: 1
     Key_name: dept_name_idx
 Seq_in_index: 1
  Column_name: name
    Collation: A
  Cardinality: 3
     Sub_part: NULL
       Packed: NULL
         Null:
   Index_type: BTREE
      Comment:
Index_Comment:
2 rows in set (0.01 sec)
```

The output shows that there are two indexes on the `department` table: one on the `dept_id` column called `PRIMARY`, and the other on the `name` column called `dept_name_idx`. Since I have created only one index so far (`dept_name_idx`), you might be wondering where the other came from; when the `department` table was created, the `create table` statement included a constraint naming the `dept_id` column as the primary key for the table. Here's the statement used to create the table:

```
CREATE TABLE department
 (dept_id SMALLINT UNSIGNED NOT NULL AUTO_INCREMENT,
  name VARCHAR(20) NOT NULL,
  CONSTRAINT pk_department PRIMARY KEY (dept_id)  );
```

When the table was created, the MySQL server automatically generated an index on the primary key column, which, in this case, is `dept_id`, and gave the index the name `PRIMARY`. I cover constraints later in this chapter.

If, after creating an index, you decide that the index is not proving useful, you can remove it via the following:

```
mysql> ALTER TABLE department
    -> DROP INDEX dept_name_idx;
Query OK, 3 rows affected (0.02 sec)
Records: 3  Duplicates: 0  Warnings: 0
```

 SQL Server and Oracle Database users must use the `drop index` command to remove an index, as in:

```
DROP INDEX dept_name_idx; (Oracle)
```

```
DROP INDEX dept_name_idx ON department (SQL Server)
```

MySQL now also supports a `drop index` command.

Unique indexes

When designing a database, it is important to consider which columns are allowed to contain duplicate data and which are not. For example, it is allowable to have two customers named John Smith in the `individual` table since each row will have a different identifier (`cust_id`), birth date, and tax number (`customer.fed_id`) to help tell them apart. You would not, however, want to allow two departments with the same name in the `department` table. You can enforce a rule against duplicate department names by creating a *unique index* on the `department.name` column.

A unique index plays multiple roles in that, along with providing all the benefits of a regular index, it also serves as a mechanism for disallowing duplicate values in the indexed column. Whenever a row is inserted or when the indexed column is modified, the database server checks the unique index to see whether the value already exists in another row in the table. Here's how you would create a unique index on the `department.name` column:

```
mysql> ALTER TABLE department
    -> ADD UNIQUE dept_name_idx (name);
Query OK, 3 rows affected (0.04 sec)
Records: 3  Duplicates: 0  Warnings: 0
```

SQL Server and Oracle Database users need only add the **unique** keyword when creating an index, as in:

```
CREATE UNIQUE INDEX dept_name_idx
ON department (name);
```

With the index in place, you will receive an error if you try to add another department with the name `'Operations'`:

```
mysql> INSERT INTO department (dept_id, name)
    -> VALUES (999, 'Operations');
ERROR 1062 (23000): Duplicate entry 'Operations' for key 'dept_name_idx'
```

You should not build unique indexes on your primary key column(s), since the server already checks uniqueness for primary key values. You may, however, create more than one unique index on the same table if you feel that it is warranted.

Multicolumn indexes

Along with the single-column indexes demonstrated thus far, you may build indexes that span multiple columns. If, for example, you find yourself searching for employees by first *and* last names, you can build an index on *both* columns together, as in:

```
mysql> ALTER TABLE employee
    -> ADD INDEX emp_names_idx (lname, fname);
Query OK, 18 rows affected (0.10 sec)
Records: 18  Duplicates: 0  Warnings: 0
```

This index will be useful for queries that specify the first and last names or just the last name, but you cannot use it for queries that specify only the employee's first name. To understand why, consider how you would find a person's phone number; if you know the person's first and last names, you can use a phone book to find the number quickly, since a phone book is organized by last name and then by first name. If you know only the person's first name, you would need to scan every entry in the phone book to find all the entries with the specified first name.

When building multiple-column indexes, therefore, you should think carefully about which column to list first, which column to list second, and so on so that the index is as useful as possible. Keep in mind, however, that there is nothing stopping you from building multiple indexes using the same set of columns but in a different order if you feel that it is needed to ensure adequate response time.

Types of Indexes

Indexing is a powerful tool, but since there are many different types of data, a single indexing strategy doesn't always do the job. The following sections illustrate the different types of indexing available from various servers.

B-tree indexes

All the indexes shown thus far are *balanced-tree indexes*, which are more commonly known as *B-tree indexes*. MySQL, Oracle Database, and SQL Server all default to B-tree indexing, so you will get a B-tree index unless you explicitly ask for another type. As you might expect, B-tree indexes are organized as trees, with one or more levels of *branch nodes* leading to a single level of *leaf nodes*. Branch nodes are used for navigating the tree, while leaf nodes hold the actual values and location information. For example, a B-tree index built on the `employee.lname` column might look something like Figure 13-1.

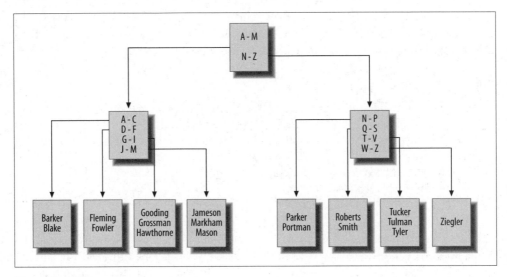

Figure 13-1. B-tree example

If you were to issue a query to retrieve all employees whose last name starts with *G*, the server would look at the top branch node (called the *root node*) and follow the link to the branch node that handles last names beginning with *A* through *M*. This branch node would, in turn, direct the server to a leaf node containing last names beginning with *G* through *I*. The server then starts reading the values in the leaf node until it encounters a value that doesn't begin with *G* (which, in this case, is `'Hawthorne'`).

As rows are inserted, updated, and deleted from the `employee` table, the server will attempt to keep the tree balanced so that there aren't far more branch/leaf nodes on one side of the root node than the other. The server can add or remove branch nodes to redistribute the values more evenly and can even add or remove an entire level of branch nodes. By keeping the tree balanced, the server is able to traverse quickly to the leaf nodes to find the desired values without having to navigate through many levels of branch nodes.

Bitmap indexes

Although B-tree indexes are great at handling columns that contain many different values, such as a customer's first/last names, they can become unwieldy when built on a column that allows only a small number of values. For example, you may decide to generate an index on the `account.product_cd` column so that you can quickly retrieve all accounts of a specific type (e.g., checking, savings). Because there are only eight different products, however, and because some products are far more popular than others, it can be difficult to maintain a balanced B-tree index as the number of accounts grows.

For columns that contain only a small number of values across a large number of rows (known as *low-cardinality* data), a different indexing strategy is needed. To handle this situation more efficiently, Oracle Database includes *bitmap indexes*, which generate a bitmap for each value stored in the column. Figure 13-2 shows what a bitmap index might look like for data in the `account.product_cd` column.

Value/row	1	2	3	4	5	6	7	8	9	10	11	12	13	14	15	16	17	18	19	20	21	22	23	24
BUS	0	0	0	0	0	0	0	0	0	0	0	0	0	0	0	0	0	0	0	0	0	1	1	0
CD	0	0	1	0	0	0	0	0	0	0	0	0	1	1	0	0	0	0	1	0	0	0	0	0
CHK	1	0	0	1	0	1	0	1	0	0	1	1	0	0	1	0	1	0	0	1	0	0	1	0
MM	0	0	0	0	0	0	1	0	0	1	0	0	0	0	0	0	0	1	0	0	0	0	0	0
SAV	0	1	0	0	1	0	0	0	1	0	0	0	0	0	0	1	0	0	0	0	0	0	0	0
SBL	0	0	0	0	0	0	0	0	0	0	0	0	0	0	0	0	0	0	0	0	0	0	0	1

Figure 13-2. Bitmap example

The index contains six bitmaps, one for each value in the `product_cd` column (two of the eight available products are not being used), and each bitmap includes a `0`/`1` value for each of the 24 rows in the `account` table. Thus, if you ask the server to retrieve all money market accounts (`product_cd = 'MM'`), the server simply finds all the `1` values in the `MM` bitmap and returns rows 7, 10, and 18. The server can also combine bitmaps if you are looking for multiple values; for example, if you want to retrieve all money market *and* savings accounts (`product_cd = 'MM'` or `product_cd = 'SAV'`), the server can perform an `OR` operation on the `MM` and `SAV` bitmaps and return rows 2, 5, 7, 9, 10, 16, and 18.

Bitmap indexes are a nice, compact indexing solution for low-cardinality data, but this indexing strategy breaks down if the number of values stored in the column climbs too high in relation to the number of rows (known as *high-cardinality* data), since the server would need to maintain too many bitmaps. For example, you would never build a bitmap index on your primary key column, since this represents the highest possible cardinality (a different value for every row).

Oracle users can generate bitmap indexes by simply adding the `bitmap` keyword to the `create index` statement, as in:

```
CREATE BITMAP INDEX acc_prod_idx ON account (product_cd);
```

Bitmap indexes are commonly used in data warehousing environments, where large amounts of data are generally indexed on columns containing relatively few values (e.g., sales quarters, geographic regions, products, salespeople).

Text indexes

If your database stores documents, you may need to allow users to search for words or phrases in the documents. You certainly don't want the server to open each document and scan for the desired text each time a search is requested, but traditional indexing strategies don't work for this situation. To handle this situation, MySQL, SQL Server, and Oracle Database include specialized indexing and search mechanisms for documents; both SQL Server and MySQL include what they call *full-text* indexes (for MySQL, full-text indexes are available only with its MyISAM storage engine), and Oracle Database includes a powerful set of tools known as *Oracle Text*. Document searches are specialized enough that I refrain from showing an example, but I wanted you to at least know what is available.

How Indexes Are Used

Indexes are generally used by the server to quickly locate rows in a particular table, after which the server visits the associated table to extract the additional information requested by the user. Consider the following query:

```
mysql> SELECT emp_id, fname, lname
    -> FROM employee
    -> WHERE emp_id IN (1, 3, 9, 15);
+--------+---------+----------+
| emp_id | fname   | lname    |
+--------+---------+----------+
|      1 | Michael | Smith    |
|      3 | Robert  | Tyler    |
|      9 | Jane    | Grossman |
|     15 | Frank   | Portman  |
+--------+---------+----------+
4 rows in set (0.00 sec)
```

For this query, the server can use the primary key index on the `emp_id` column to locate employee IDs 1, 3, 9, and 15 in the `employee` table, and then visit each of the four rows to retrieve the first and last name columns.

If the index contains everything needed to satisfy the query, however, the server doesn't need to visit the associated table. To illustrate, let's look at how the query optimizer approaches the same query with different indexes in place.

The query, which aggregates account balances for specific customers, looks as follows:

```
mysql> SELECT cust_id, SUM(avail_balance) tot_bal
    -> FROM account
    -> WHERE cust_id IN (1, 5, 9, 11)
    -> GROUP BY cust_id;
+---------+----------+
| cust_id | tot_bal  |
+---------+----------+
|       1 |  4557.75 |
|       5 |  2237.97 |
|       9 | 10971.22 |
|      11 |  9345.55 |
+---------+----------+
4 rows in set (0.00 sec)
```

To see how MySQL's query optimizer decides to execute the query, I use the `explain` statement to ask the server to show the execution plan for the query rather than executing the query:

```
mysql> EXPLAIN SELECT cust_id, SUM(avail_balance) tot_bal
    -> FROM account
    -> WHERE cust_id IN (1, 5, 9, 11)
    -> GROUP BY cust_id \G
*************************** 1. row ***************************
           id: 1
  select_type: SIMPLE
        table: account
         type: index
possible_keys: fk_a_cust_id
          key: fk_a_cust_id
      key_len: 4
          ref: NULL
         rows: 24
        Extra: Using where
1 row in set (0.00 sec)
```

 Each database server includes tools to allow you to see how the query optimizer handles your SQL statement. SQL Server allows you to see an execution plan by issuing the statement `set showplan_text on` before running your SQL statement. Oracle Database includes the `explain plan` statement, which writes the execution plan to a special table called `plan_table`.

Without going into too much detail, here's what the execution plan tells you:

- The fk_a_cust_id index is used to find the rows in the account table that satisfy the where clause.
- After reading the index, the server expects to read all 24 rows of the account table to gather the available balance data, since it doesn't know that there might be other customers besides IDs 1, 5, 9, and 11.

The fk_a_cust_id index is another index generated automatically by the server, but this time it is because of a foreign key constraint rather than a primary key constraint (more on this later in the chapter). The fk_a_cust_id index is built on the account.cust_id column, so the server is using the index to locate customer IDs 1, 5, 9, and 11 in the account table and is then visiting those rows to retrieve and aggregate the available balance data.

Next, I will add a new index called acc_bal_idx on both the cust_id and avail_balance columns:

```
mysql> ALTER TABLE account
    -> ADD INDEX acc_bal_idx (cust_id, avail_balance);
Query OK, 24 rows affected (0.03 sec)
Records: 24  Duplicates: 0  Warnings: 0
```

With this index in place, let's see how the query optimizer approaches the same query:

```
mysql> EXPLAIN SELECT cust_id, SUM(avail_balance) tot_bal
    -> FROM account
    -> WHERE cust_id IN (1, 5, 9, 11)
    -> GROUP BY cust_id \G
*************************** 1. row ***************************
           id: 1
  select_type: SIMPLE
        table: account
         type: range
possible_keys: acc_bal_idx
          key: acc_bal_idx
      key_len: 4
          ref: NULL
         rows: 8
        Extra: Using where; Using index
1 row in set (0.01 sec)
```

Comparing the two execution plans yields the following differences:

- The optimizer is using the new acc_bal_idx index instead of the fk_a_cust_id index.
- The optimizer anticipates needing only eight rows instead of 24.
- The account table is not needed (designated by Using index in the Extra column) to satisfy the query results.

Therefore, the server can use indexes to help locate rows in the associated table, or the server can use an index as though it were a table as long as the index contains all the columns needed by the query.

The process that I just led you through is an example of query tuning. Tuning involves looking at an SQL statement and determining the resources available to the server to execute the statement. You can decide to modify the SQL statement, to adjust the database resources, or to do both in order to make a statement run more efficiently. Tuning is a detailed topic, and I strongly urge you to either read your server's tuning guide or pick up a good tuning book so that you can see all the different approaches available for your server.

The Downside of Indexes

If indexes are so great, why not index everything? Well, the key to understanding why more indexes are not necessarily a good thing is to keep in mind that every index is a table (a special type of table, but still a table). Therefore, every time a row is added to or removed from a table, all indexes on that table must be modified. When a row is updated, any indexes on the column or columns that were affected need to be modified as well. Therefore, the more indexes you have, the more work the server needs to do to keep all schema objects up-to-date, which tends to slow things down.

Indexes also require disk space as well as some amount of care from your administrators, so the best strategy is to add an index when a clear need arises. If you need an index for only special purposes, such as a monthly maintenance routine, you can always add the index, run the routine, and then drop the index until you need it again. In the case of data warehouses, where indexes are crucial during business hours as users run reports and ad hoc queries but are problematic when data is being loaded into the warehouse overnight, it is a common practice to drop the indexes before data is loaded and then re-create them before the warehouse opens for business.

In general, you should strive to have neither too many indexes nor too few. If you aren't sure how many indexes you should have, you can use this strategy as a default:

- Make sure all primary key columns are indexed (most servers automatically create unique indexes when you create primary key constraints). For multicolumn primary keys, consider building additional indexes on a subset of the primary key columns, or on all the primary key columns but in a different order than the primary key constraint definition.

- Build indexes on all columns that are referenced in foreign key constraints. Keep in mind that the server checks to make sure there are no child rows when a parent is deleted, so it must issue a query to search for a particular value in the column. If there's no index on the column, the entire table must be scanned.

- Index any columns that will frequently be used to retrieve data. Most date columns are good candidates, along with short (3- to 50-character) string columns.

After you have built your initial set of indexes, try to capture actual queries against your tables, and modify your indexing strategy to fit the most-common access paths.

Constraints

A constraint is simply a restriction placed on one or more columns of a table. There are several different types of constraints, including:

Primary key constraints
    Identify the column or columns that guarantee uniqueness within a table

Foreign key constraints
    Restrict one or more columns to contain only values found in another table's primary key columns, and may also restrict the allowable values in other tables if `update cascade` or `delete cascade` rules are established

Unique constraints
    Restrict one or more columns to contain unique values within a table (primary key constraints are a special type of unique constraint)

Check constraints
    Restrict the allowable values for a column

Without constraints, a database's consistency is suspect. For example, if the server allows you to change a customer's ID in the `customer` table without changing the same customer ID in the `account` table, then you will end up with accounts that no longer point to valid customer records (known as *orphaned rows*). With primary and foreign key constraints in place, however, the server will either raise an error if an attempt is made to modify or delete data that is referenced by other tables, or propagate the changes to other tables for you (more on this shortly).

 If you want to use foreign key constraints with the MySQL server, you must use the InnoDB storage engine for your tables. Foreign key constraints are not supported in the Falcon engine as of version 6.0.4, but they will be supported in later versions.

Constraint Creation

Constraints are generally created at the same time as the associated table via the `create table` statement. To illustrate, here's an example from the schema generation script for this book's example database:

```
CREATE TABLE product
 (product_cd VARCHAR(10) NOT NULL,
  name VARCHAR(50) NOT NULL,
  product_type_cd VARCHAR (10) NOT NULL,
  date_offered DATE,
  date_retired DATE,
      CONSTRAINT fk_product_type_cd FOREIGN KEY (product_type_cd)
        REFERENCES product_type (product_type_cd),
      CONSTRAINT pk_product PRIMARY KEY (product_cd)
 );
```

The product table includes two constraints: one to specify that the product_cd column serves as the primary key for the table, and another to specify that the product_type_cd column serves as a foreign key to the product_type table. Alternatively, you can create the product table without constraints, and add the primary and foreign key constraints later via alter table statements:

```
ALTER TABLE product
ADD CONSTRAINT pk_product PRIMARY KEY (product_cd);

ALTER TABLE product
ADD CONSTRAINT fk_product_type_cd FOREIGN KEY (product_type_cd)
  REFERENCES product_type (product_type_cd);
```

If you want to remove a primary or foreign key constraint, you can use the alter table statement again, except that you specify drop instead of add, as in:

```
ALTER TABLE product
DROP PRIMARY KEY;

ALTER TABLE product
DROP FOREIGN KEY fk_product_type_cd;
```

While it is unusual to drop a primary key constraint, foreign key constraints are sometimes dropped during certain maintenance operations and then reestablished.

Constraints and Indexes

As you saw earlier in the chapter, constraint creation sometimes involves the automatic generation of an index. However, database servers behave differently regarding the relationship between constraints and indexes. Table 13-1 shows how MySQL, SQL Server, and Oracle Database handle the relationship between constraints and indexes.

Table 13-1. Constraint generation

Constraint type	MySQL	SQL Server	Oracle Database
Primary key constraints	Generates unique index	Generates unique index	Uses existing index or creates new index
Foreign key constraints	Generates index	Does not generate index	Does not generate index
Unique constraints	Generates unique index	Generates unique index	Uses existing index or creates new index

MySQL, therefore, generates a new index to enforce primary key, foreign key, and unique constraints, SQL Server generates a new index for primary key and unique constraints but *not* for foreign key constraints, and Oracle Database takes the same approach as SQL Server except that Oracle will use an existing index (if an appropriate one exists) to enforce primary key and unique constraints. Although neither SQL Server nor Oracle Database generates an index for a foreign key constraint, both servers' documentation advises that indexes be created for every foreign key.

Cascading Constraints

With foreign key constraints in place, if a user attempts to insert a new row or change an existing row such that a foreign key column doesn't have a matching value in the parent table, the server raises an error. To illustrate, here's a look at the data in the product and product_type tables:

```
mysql> SELECT product_type_cd, name
    -> FROM product_type;
+-----------------+------------------------------+
| product_type_cd | name                         |
+-----------------+------------------------------+
| ACCOUNT         | Customer Accounts            |
| INSURANCE       | Insurance Offerings          |
| LOAN            | Individual and Business Loans |
+-----------------+------------------------------+
3 rows in set (0.00 sec)

mysql> SELECT product_type_cd, product_cd, name
    -> FROM product
    -> ORDER BY product_type_cd;
+-----------------+------------+-----------------------+
| product_type_cd | product_cd | name                  |
+-----------------+------------+-----------------------+
| ACCOUNT         | CD         | certificate of deposit |
| ACCOUNT         | CHK        | checking account      |
| ACCOUNT         | MM         | money market account  |
| ACCOUNT         | SAV        | savings account       |
| LOAN            | AUT        | auto loan             |
| LOAN            | BUS        | business line of credit |
| LOAN            | MRT        | home mortgage         |
| LOAN            | SBL        | small business loan   |
+-----------------+------------+-----------------------+
8 rows in set (0.01 sec)
```

There are three different values for the product_type_cd column in the product_type table (ACCOUNT, INSURANCE, and LOAN). Of the three values, two of them (ACCOUNT and LOAN) are referenced in the product table's product_type_cd column.

The following statement attempts to change the product_type_cd column in the product table to a value that doesn't exist in the product_type table:

```
mysql> UPDATE product
    -> SET product_type_cd = 'XYZ'
    -> WHERE product_type_cd = 'LOAN';
ERROR 1452 (23000): Cannot add or update a child row: a foreign key constraint
fails ('bank'.'product', CONSTRAINT 'fk_product_type_cd' FOREIGN KEY
('product_type_cd') REFERENCES 'product_type' ('product_type_cd'))
```

Because of the foreign key constraint on the product.product_type_cd column, the server does not allow the update to succeed, since there is no row in the product_type table with a value of XYZ in the product_type_cd column. Thus, the foreign

---

key constraint doesn't let you change a child row if there is no corresponding value in the parent.

What would happen, however, if you tried to change the *parent* row in the product_type table to XYZ? Here's an update statement that attempts to change the LOAN product type to XYZ:

```
mysql> UPDATE product_type
    -> SET product_type_cd = 'XYZ'
    -> WHERE product_type_cd = 'LOAN';
ERROR 1451 (23000): Cannot delete or update a parent row: a foreign key
constraint fails ('bank'.'product', CONSTRAINT 'fk_product_type_cd' FOREIGN KEY
('product_type_cd') REFERENCES 'product_type' ('product_type_cd'))
```

Once again, an error is raised; this time because there are child rows in the product table whose product_type_cd column contains the value LOAN. This is the default behavior for foreign key constraints, but it is not the only possible behavior; instead, you can instruct the server to propagate the change to all child rows for you, thus preserving the integrity of the data. Known as a *cascading update*, this variation of the foreign key constraint can be installed by removing the existing foreign key and adding a new one that includes the on update cascade clause:

```
mysql> ALTER TABLE product
    -> DROP FOREIGN KEY fk_product_type_cd;
Query OK, 8 rows affected (0.02 sec)
Records: 8  Duplicates: 0  Warnings: 0

mysql> ALTER TABLE product
    -> ADD CONSTRAINT fk_product_type_cd FOREIGN KEY (product_type_cd)
    ->   REFERENCES product_type (product_type_cd)
    ->   ON UPDATE CASCADE;
Query OK, 8 rows affected (0.03 sec)
Records: 8  Duplicates: 0  Warnings: 0
```

With this modified constraint in place, let's see what happens when the previous update statement is attempted again:

```
mysql> UPDATE product_type
    -> SET product_type_cd = 'XYZ'
    -> WHERE product_type_cd = 'LOAN';
Query OK, 1 row affected (0.01 sec)
Rows matched: 1  Changed: 1  Warnings: 0
```

This time, the statement succeeds. To verify that the change was propagated to the product table, here's another look at the data in both tables:

```
mysql> SELECT product_type_cd, name
    -> FROM product_type;
+-----------------+------------------------------+
| product_type_cd | name                         |
+-----------------+------------------------------+
| ACCOUNT         | Customer Accounts            |
| INSURANCE       | Insurance Offerings          |
| XYZ             | Individual and Business Loans |
```

```
+-----------------+-----------------------------+
3 rows in set (0.02 sec)

mysql> SELECT product_type_cd, product_cd, name
    -> FROM product
    -> ORDER BY product_type_cd;
+-----------------+------------+-------------------------+
| product_type_cd | product_cd | name                    |
+-----------------+------------+-------------------------+
| ACCOUNT         | CD         | certificate of deposit  |
| ACCOUNT         | CHK        | checking account        |
| ACCOUNT         | MM         | money market account    |
| ACCOUNT         | SAV        | savings account         |
| XYZ             | AUT        | auto loan               |
| XYZ             | BUS        | business line of credit |
| XYZ             | MRT        | home mortgage           |
| XYZ             | SBL        | small business loan     |
+-----------------+------------+-------------------------+
8 rows in set (0.01 sec)
```

As you can see, the change to the product_type table has been propagated to the product table as well. Along with cascading updates, you can specify *cascading deletes* as well. A cascading delete *removes* rows from the child table when a row is deleted from the parent table. To specify cascading deletes, use the on delete cascade clause, as in:

```
ALTER TABLE product
ADD CONSTRAINT fk_product_type_cd FOREIGN KEY (product_type_cd)
  REFERENCES product_type (product_type_cd)
  ON UPDATE CASCADE
  ON DELETE CASCADE;
```

With this version of the constraint in place, the server will now update child rows in the product table when a row in the product_type table is updated, as well as delete child rows in the product table when a row in the product_type table is deleted.

Cascading constraints are one case in which constraints *do* directly affect the code that you write. You need to know which constraints in your database specify cascading updates and/or deletes so that you know the full effect of your update and delete statements.

Test Your Knowledge

Work through the following exercises to test your knowledge of indexes and constraints. When you're done, compare your solutions with those in Appendix C.

Exercise 13-1

Modify the `account` table so that customers may not have more than one account for each product.

Exercise 13-2

Generate a multicolumn index on the `transaction` table that could be used by both of the following queries:

```
SELECT txn_date, account_id, txn_type_cd, amount
FROM transaction
WHERE txn_date > cast('2008-12-31 23:59:59' as datetime);

SELECT txn_date, account_id, txn_type_cd, amount
FROM transaction
WHERE txn_date > cast('2008-12-31 23:59:59' as datetime)
  AND amount < 1000;
```

CHAPTER 14

Views

Well-designed applications generally expose a public interface while keeping implementation details private, thereby enabling future design changes without impacting end users. When designing your database, you can achieve a similar result by keeping your tables private and allowing your users to access data only through a set of *views*. This chapter strives to define what views are, how they are created, and when and how you might want to use them.

What Are Views?

A view is simply a mechanism for querying data. Unlike tables, views do not involve data storage; you won't need to worry about views filling up your disk space. You create a view by assigning a name to a `select` statement, and then storing the query for others to use. Other users can then use your view to access data just as though they were querying tables directly (in fact, they may not even know they are using a view).

As a simple example, let's say that you want to partially obscure the federal IDs (Social Security numbers and corporate identifiers) in the `customer` table. The customer service department, for example, may need access to just the last portion of the federal ID in order to verify the identity of a caller, but exposing the entire number would violate the company's privacy policy. Therefore, instead of allowing direct access to the `customer` table, you define a view called `customer_vw` and mandate that all bank personnel use it to access customer data. Here's the view definition:

```
CREATE VIEW customer_vw
  (cust_id,
   fed_id,
   cust_type_cd,
   address,
   city,
   state,
   zipcode
  )
AS
SELECT cust_id,
```

```
    concat('ends in ', substr(fed_id, 8, 4)) fed_id,
    cust_type_cd,
    address,
    city,
    state,
    postal_code
FROM customer;
```

The first part of the statement lists the view's column names, which may be different from those of the underlying table (e.g., the `customer_vw` view has a column named `zipcode` which maps to the `customer.postal_code` column). The second part of the statement is a `select` statement, which must contain one expression for each column in the view.

When the `create view` statement is executed, the database server simply stores the view definition for future use; the query is not executed, and no data is retrieved or stored. Once the view has been created, users can query it just like they would a table, as in:

```
mysql> SELECT cust_id, fed_id, cust_type_cd
    -> FROM customer_vw;
+---------+--------------+--------------+
| cust_id | fed_id       | cust_type_cd |
+---------+--------------+--------------+
|       1 | ends in 1111 | I            |
|       2 | ends in 2222 | I            |
|       3 | ends in 3333 | I            |
|       4 | ends in 4444 | I            |
|       5 | ends in 5555 | I            |
|       6 | ends in 6666 | I            |
|       7 | ends in 7777 | I            |
|       8 | ends in 8888 | I            |
|       9 | ends in 9999 | I            |
|      10 | ends in 111  | B            |
|      11 | ends in 222  | B            |
|      12 | ends in 333  | B            |
|      13 | ends in 444  | B            |
+---------+--------------+--------------+
13 rows in set (0.02 sec)
```

The actual query that the server executes is neither the one submitted by the user nor the query attached to the view definition. Instead, the server merges the two together to create another statement, which in this case looks as follows:

```
SELECT cust_id,
    concat('ends in ', substr(fed_id, 8, 4)) fed_id,
    cust_type_cd
FROM customer;
```

Even though the `customer_vw` view definition includes seven columns of the `customer` table, the query executed by the server retrieves only three of the seven. As you'll see later in the chapter, this is an important distinction if some of the columns in your view are attached to functions or subqueries.

From the user's standpoint, a view looks exactly like a table. If you want to know what columns are available in a view, you can use MySQL's (or Oracle's) `describe` command to examine it:

```
mysql> describe customer_vw;
+--------------+-------------------+------+-----+---------+-------+
| Field        | Type              | Null | Key | Default | Extra |
+--------------+-------------------+------+-----+---------+-------+
| cust_id      | int(10) unsigned  | NO   |     | 0       |       |
| fed_id       | varchar(12)       | YES  |     | NULL    |       |
| cust_type_cd | enum('I','B')     | NO   |     | NULL    |       |
| address      | varchar(30)       | YES  |     | NULL    |       |
| city         | varchar(20)       | YES  |     | NULL    |       |
| state        | varchar(20)       | YES  |     | NULL    |       |
| postal_code  | varchar(10)       | YES  |     | NULL    |       |
+--------------+-------------------+------+-----+---------+-------+
7 rows in set (1.40 sec)
```

You are free to use any clauses of the `select` statement when querying through a view, including `group by`, `having`, and `order by`. Here's an example:

```
mysql> SELECT cust_type_cd, count(*)
    -> FROM customer_vw
    -> WHERE state = 'MA'
    -> GROUP BY cust_type_cd
    -> ORDER BY 1;
+--------------+----------+
| cust_type_cd | count(*) |
+--------------+----------+
| I            |        7 |
| B            |        2 |
+--------------+----------+
2 rows in set (0.22 sec)
```

In addition, you can join views to other tables (or even to other views) within a query, as in:

```
mysql> SELECT cst.cust_id, cst.fed_id, bus.name
    -> FROM customer_vw cst INNER JOIN business bus
    ->   ON cst.cust_id = bus.cust_id;
+---------+-------------+-----------------------+
| cust_id | fed_id      | name                  |
+---------+-------------+-----------------------+
|      10 | ends in 111 | Chilton Engineering   |
|      11 | ends in 222 | Northeast Cooling Inc.|
|      12 | ends in 333 | Superior Auto Body    |
|      13 | ends in 444 | AAA Insurance Inc.    |
+---------+-------------+-----------------------+
4 rows in set (0.24 sec)
```

This query joins the `customer_vw` view to the `business` table in order to retrieve only business customers.

Why Use Views?

In the previous section, I demonstrated a simple view whose sole purpose was to mask the contents of the `customer.fed_id` column. While views are often employed for this purpose, there are many reasons for using views, as I demonstrate in the following subsections.

Data Security

If you create a table and allow users to query it, they will be able to access every column and every row in the table. As I pointed out earlier, however, your table may include some columns that contain sensitive data, such as identification numbers or credit card numbers; not only is it a bad idea to expose such data to all users, but also it might violate your company's privacy policies, or even state or federal laws, to do so.

The best approach for these situations is to keep the table private (i.e., don't grant `select` permission to any users) and then to create one or more views that either omit or obscure (such as the `'ends in ####'` approach taken with the `customer_vw.fed_id` column) the sensitive columns. You may also constrain which *rows* a set of users may access by adding a `where` clause to your view definition. For example, the next view definition allows only business customers to be queried:

```
CREATE VIEW business_customer_vw
 (cust_id,
  fed_id,
  cust_type_cd,
  address,
  city,
  state,
  zipcode
 )
AS
SELECT cust_id,
  concat('ends in ', substr(fed_id, 8, 4)) fed_id,
  cust_type_cd,
  address,
  city,
  state,
  postal_code
FROM customer
WHERE cust_type_cd = 'B'
```

If you provide this view to your corporate banking department, they will be able to access only business accounts because the condition in the view's `where` clause will always be included in their queries.

Oracle Database users have another option for securing both rows and columns of a table: Virtual Private Database (VPD). VPD allows you to attach policies to your tables, after which the server will modify a user's query as necessary to enforce the policies. For example, if you enact a policy that members of the corporate banking department can see only business accounts, then the condition cust_type_cd = 'B' will be added to all of their queries against the customer table.

Data Aggregation

Reporting applications generally require aggregated data, and views are a great way to make it appear as though data is being pre-aggregated and stored in the database. As an example, let's say that an application generates a report each month showing the number of accounts and total deposits for every customer. Rather than allowing the application developers to write queries against the base tables, you could provide them with the following view:

```
CREATE VIEW customer_totals_vw
 (cust_id,
  cust_type_cd,
  cust_name,
  num_accounts,
  tot_deposits
 )
AS
SELECT cst.cust_id, cst.cust_type_cd,
  CASE
    WHEN cst.cust_type_cd = 'B' THEN
      (SELECT bus.name FROM business bus WHERE bus.cust_id = cst.cust_id)
    ELSE
      (SELECT concat(ind.fname, ' ', ind.lname)
       FROM individual ind
       WHERE ind.cust_id = cst.cust_id)
  END cust_name,
  sum(CASE WHEN act.status = 'ACTIVE' THEN 1 ELSE 0 END) tot_active_accounts,
  sum(CASE WHEN act.status = 'ACTIVE' THEN act.avail_balance ELSE 0 END) tot_balance
FROM customer cst INNER JOIN account act
  ON act.cust_id = cst.cust_id
GROUP BY cst.cust_id, cst.cust_type_cd;
```

Using this approach gives you a great deal of flexibility as a database designer. If you decide at some point in the future that query performance would improve dramatically if the data were preaggregated in a table rather than summed using a view, you can create a customer_totals table and modify the customer_totals_vw view definition to retrieve data from this table. Before modifying the view definition, you can use it to populate the new table. Here are the necessary SQL statements for this scenario:

```
mysql> CREATE TABLE customer_totals
    -> AS
    -> SELECT * FROM customer_totals_vw;
Query OK, 13 rows affected (3.33 sec)
```

```
Records: 13  Duplicates: 0  Warnings: 0

mysql> CREATE OR REPLACE VIEW customer_totals_vw
    -> (cust_id,
    ->  cust_type_cd,
    ->  cust_name,
    ->  num_accounts,
    ->  tot_deposits
    -> )
    -> AS
    -> SELECT cust_id,  cust_type_cd,  cust_name,  num_accounts,  tot_deposits
    -> FROM customer_totals;
Query OK, 0 rows affected (0.02 sec)
```

From now on, all queries that use the customer_totals_vw view will pull data from the new customer_totals table, meaning that users will see a performance improvement without needing to modify their queries.

Hiding Complexity

One of the most common reasons for deploying views is to shield end users from complexity. For example, let's say that a report is created each month showing the number of employees, the total number of active accounts, and the total number of transactions for each branch. Rather than expecting the report designer to navigate four different tables to gather the necessary data, you could provide a view that looks as follows:

```
CREATE VIEW branch_activity_vw
 (branch_name,
  city,
  state,
  num_employees,
  num_active_accounts,
  tot_transactions
 )
AS
SELECT br.name, br.city, br.state,
 (SELECT count(*)
  FROM employee emp
  WHERE emp.assigned_branch_id = br.branch_id) num_emps,
 (SELECT count(*)
  FROM account acnt
  WHERE acnt.status = 'ACTIVE' AND acnt.open_branch_id = br.branch_id) num_accounts,
 (SELECT count(*)
  FROM transaction txn
  WHERE txn.execution_branch_id = br.branch_id) num_txns
FROM branch br;
```

This view definition is interesting because three of the six column values are generated using scalar subqueries. If someone uses this view but does *not* reference the num_employees, num_active_accounts, or tot_transactions column, then none of the subqueries will be executed.

Joining Partitioned Data

Some database designs break large tables into multiple pieces in order to improve performance. For example, if the `transaction` table became large, the designers may decide to break it into two tables: `transaction_current`, which holds the latest six months' of data, and `transaction_historic`, which holds all data up to six months ago. If a customer wants to see all the transactions for a particular account, you would need to query both tables. By creating a view that queries both tables and combines the results together, however, you can make it look like all transaction data is stored in a single table. Here's the view definition:

```
CREATE VIEW transaction_vw
 (txn_date,
  account_id,
  txn_type_cd,
  amount,
  teller_emp_id,
  execution_branch_id,
  funds_avail_date
 )
AS
SELECT txn_date,  account_id,  txn_type_cd,  amount,  teller_emp_id,
  execution_branch_id,  funds_avail_date
FROM transaction_historic
UNION ALL
SELECT txn_date,  account_id,  txn_type_cd,  amount,  teller_emp_id,
  execution_branch_id,  funds_avail_date
FROM transaction_current;
```

Using a view in this case is a good idea because it allows the designers to change the structure of the underlying data without the need to force all database users to modify their queries.

Updatable Views

If you provide users with a set of views to use for data retrieval, what should you do if the users also need to modify the same data? It might seem a bit strange, for example, to force the users to retrieve data using a view, but then allow them to directly modify the underlying table using `update` or `insert` statements. For this purpose, MySQL, Oracle Database, and SQL Server all allow you to modify data through a view, as long as you abide by certain restrictions. In the case of MySQL, a view is updatable if the following conditions are met:

- No aggregate functions are used (`max()`, `min()`, `avg()`, etc.).
- The view does not employ `group by` or `having` clauses.
- No subqueries exist in the `select` or `from` clause, and any subqueries in the `where` clause do not refer to tables in the `from` clause.
- The view does not utilize `union`, `union all`, or `distinct`.

- The from clause includes at least one table or updatable view.
- The from clause uses only inner joins if there is more than one table or view.

To demonstrate the utility of updatable views, it might be best to start with a simple view definition and then to move to a more complex view.

Updating Simple Views

The view at the beginning of the chapter is about as simple as it gets, so let's start there:

```
CREATE VIEW customer_vw
 (cust_id,
  fed_id,
  cust_type_cd,
  address,
  city,
  state,
  zipcode
 )
AS
SELECT cust_id,
  concat('ends in ', substr(fed_id, 8, 4)) fed_id,
  cust_type_cd,
  address,
  city,
  state,
  postal_code
FROM customer;
```

The customer_vw view queries a single table, and only one of the seven columns is derived via an expression. This view definition doesn't violate any of the restrictions listed earlier, so you can use it to modify data in the customer table, as in:

```
mysql> UPDATE customer_vw
    -> SET city = 'Woooburn'
    -> WHERE city = 'Woburn';
Query OK, 1 row affected (0.34 sec)
Rows matched: 1  Changed: 1  Warnings: 0
```

As you can see, the statement claims to have modified one row, but let's check the underlying customer table just to be sure:

```
mysql> SELECT DISTINCT city FROM customer;
+------------+
| city       |
+------------+
| Lynnfield  |
| Woooburn   |
| Quincy     |
| Waltham    |
| Salem      |
| Wilmington |
| Newton     |
```

```
+------------+
7 rows in set (0.12 sec)
```

While you can modify most of the columns in the view in this fashion, you will not be able to modify the `fed_id` column, since it is derived from an expression:

```
mysql> UPDATE customer_vw
    -> SET city = 'Woburn', fed_id = '999999999'
    -> WHERE city = 'Woooburn';
ERROR 1348 (HY000): Column 'fed_id' is not updatable
```

In this case, it may not be a bad thing, since the whole point of the view is to obscure the federal identifiers.

If you want to insert data using the `customer_vw` view, you are out of luck; views that contain derived columns cannot be used for inserting data, even if the derived columns are not included in the statement. For example, the next statement attempts to populate only the `cust_id`, `cust_type_cd`, and `city` columns using the `customer_vw` view:

```
mysql> INSERT INTO customer_vw(cust_id, cust_type_cd, city)
    -> VALUES (9999, 'I', 'Worcester');
ERROR 1471 (HY000): The target table customer_vw of the INSERT is not insertable
-into
```

Now that you have seen the limitations of simple views, the next section will demonstrate the use of a view that joins multiple tables.

Updating Complex Views

While single-table views are certainly common, many of the views that you come across will include multiple tables in the `from` clause of the underlying query. The next view, for example, joins the `business` and `customer` tables so that all the data for business customers can be easily queried:

```
CREATE VIEW business_customer_vw
 (cust_id,
  fed_id,
  address,
  city,
  state,
  postal_code,
  business_name,
  state_id,
  incorp_date
 )
AS
SELECT cst.cust_id,
  cst.fed_id,
  cst.address,
  cst.city,
  cst.state,
  cst.postal_code,
  bsn.name,
  bsn.state_id,
```

```
      bsn.incorp_date
FROM customer cst INNER JOIN business bsn
  ON cst.cust_id = bsn.cust_id
WHERE cust_type_cd = 'B';
```

You may use this view to update data in either the customer or the business table, as
the following statements demonstrate:

```
mysql> UPDATE business_customer_vw
    -> SET postal_code = '99999'
    -> WHERE cust_id = 10;
Query OK, 1 row affected (0.09 sec)
Rows matched: 1  Changed: 1  Warnings: 0

mysql> UPDATE business_customer_vw
    -> SET incorp_date = '2008-11-17'
    -> WHERE cust_id = 10;
Query OK, 1 row affected (0.11 sec)
Rows matched: 1  Changed: 1  Warnings: 0
```

The first statement modifies the customer.postal_code column, whereas the second
statement modifies the business.incorp_date column. You might be wondering what
happens if you try to update columns from *both* tables in a single statement, so let's
find out:

```
mysql> UPDATE business_customer_vw
    -> SET postal_code = '88888', incorp_date = '2008-10-31'
    -> WHERE cust_id = 10;
ERROR 1393 (HY000): Can not modify more than one base table through a join view
'bank.business_customer_vw'
```

As you can see, you are allowed to modify both of the underlying tables, as long as you
don't try to do it with a single statement. Now let's try to *insert* data into both tables
for a new customer (cust_id = 99):

```
mysql> INSERT INTO business_customer_vw
    -> (cust_id, fed_id, address, city, state, postal_code)
    -> VALUES (99, '04-9999999', '99 Main St.', 'Peabody', 'MA', '01975');
Query OK, 1 row affected (0.07 sec)

mysql> INSERT INTO business_customer_vw
    -> (cust_id, business_name, state_id, incorp_date)
    -> VALUES (99, 'Ninety-Nine Restaurant', '99-999-999', '1999-01-01');
ERROR 1393 (HY000): Can not modify more than one base table through a join view
'bank.business_customer_vw'
```

The first statement, which attempts to insert data into the customer table, works fine,
but the second statement, which attempts to insert a row into the business table, raises
an exception. The second statement fails because both tables include a cust_id column,
but the cust_id column in the view definition is mapped to the customer.cust_id column. Therefore, it is not possible to insert data into the business table using the preceding view definition.

Oracle Database and SQL Server also allow data to be inserted and updated through views, but, like MySQL, there are many restrictions. If you are willing to write some PL/SQL or Transact-SQL, however, you can use a feature called *instead-of triggers*, which allows you to essentially intercept insert, update, and delete statements against a view, and write custom code to incorporate the changes. Without this type of feature, there are simply too many restrictions to make updating through views a feasible strategy for nontrivial applications.

Test Your Knowledge

Test your understanding of views by working through the following exercises. When you're done, compare your solutions with those in Appendix C.

Exercise 14-1

Create a view that queries the employee table and generates the following output when queried with no where clause:

```
+-----------------+------------------+
| supervisor_name | employee_name    |
+-----------------+------------------+
| NULL            | Michael Smith    |
| Michael Smith   | Susan Barker     |
| Michael Smith   | Robert Tyler     |
| Robert Tyler    | Susan Hawthorne  |
| Susan Hawthorne | John Gooding     |
| Susan Hawthorne | Helen Fleming    |
| Helen Fleming   | Chris Tucker     |
| Helen Fleming   | Sarah Parker     |
| Helen Fleming   | Jane Grossman    |
| Susan Hawthorne | Paula Roberts    |
| Paula Roberts   | Thomas Ziegler   |
| Paula Roberts   | Samantha Jameson |
| Susan Hawthorne | John Blake       |
| John Blake      | Cindy Mason      |
| John Blake      | Frank Portman    |
| Susan Hawthorne | Theresa Markham  |
| Theresa Markham | Beth Fowler      |
| Theresa Markham | Rick Tulman      |
+-----------------+------------------+
18 rows in set (1.47 sec)
```

Exercise 14-2

The bank president would like to have a report showing the name and city of each branch, along with the total balances of all accounts opened at the branch. Create a view to generate the data.

CHAPTER 15

Metadata

Along with storing all of the data that various users insert into a database, a database server also needs to store information about all of the database objects (tables, views, indexes, etc.) that were created to store this data. The database server stores this information, not surprisingly, in a database. This chapter discusses how and where this information, known as *metadata*, is stored, how you can access it, and how you can use it to build flexible systems.

Data About Data

Metadata is essentially data about data. Every time you create a database object, the database server needs to record various pieces of information. For example, if you were to create a table with multiple columns, a primary key constraint, three indexes, and a foreign key constraint, the database server would need to store all the following information:

- Table name
- Table storage information (tablespace, initial size, etc.)
- Storage engine
- Column names
- Column data types
- Default column values
- NOT NULL column constraints
- Primary key columns
- Primary key name
- Name of primary key index
- Index names
- Index types (B-tree, bitmap)
- Indexed columns

- Index column sort order (ascending or descending)
- Index storage information
- Foreign key name
- Foreign key columns
- Associated table/columns for foreign keys

This data is collectively known as the *data dictionary* or *system catalog*. The database server needs to store this data persistently, and it needs to be able to quickly retrieve this data in order to verify and execute SQL statements. Additionally, the database server must safeguard this data so that it can be modified only via an appropriate mechanism, such as the `alter table` statement.

While standards exist for the exchange of metadata between different servers, every database server uses a different mechanism to publish metadata, such as:

- A set of views, such as Oracle Database's `user_tables` and `all_constraints` views
- A set of system-stored procedures, such as SQL Server's `sp_tables` procedure or Oracle Database's `dbms_metadata` package
- A special database, such as MySQL's `information_schema` database

Along with SQL Server's system-stored procedures, which are a vestige of its Sybase lineage, SQL Server also includes a special schema called `information_schema` that is provided automatically within each database. Both MySQL and SQL Server provide this interface to conform with the ANSI SQL:2003 standard. The remainder of this chapter discusses the `information_schema` objects that are available in MySQL and SQL Server.

Information_Schema

All of the objects available within the `information_schema` database (or schema, in the case of SQL Server) are views. Unlike the `describe` utility, which I used in several chapters of this book as a way to show the structure of various tables and views, the views within `information_schema` can be queried, and, thus, used programmatically (more on this later in the chapter). Here's an example that demonstrates how to retrieve the names of all of the tables in the `bank` database:

```
mysql> SELECT table_name, table_type
    -> FROM information_schema.tables
    -> WHERE table_schema = 'bank'
    -> ORDER BY 1;
+----------------------+------------+
| table_name           | table_type |
+----------------------+------------+
| account              | BASE TABLE |
| branch               | BASE TABLE |
| branch_activity_vw   | VIEW       |
| business             | BASE TABLE |
```

```
| business_customer_vw | VIEW       |
| customer             | BASE TABLE |
| customer_vw          | VIEW       |
| department           | BASE TABLE |
| employee             | BASE TABLE |
| employee_vw          | VIEW       |
| individual           | BASE TABLE |
| nh_customer_vw       | VIEW       |
| officer              | BASE TABLE |
| product              | BASE TABLE |
| product_type         | BASE TABLE |
| transaction          | BASE TABLE |
+----------------------+------------+
16 rows in set (0.02 sec)
```

Along with the various tables we created back in Chapter 2, the results show several of the views that I demonstrated in Chapter 14. If you want to exclude the views, simply add another condition to the where clause:

```
mysql> SELECT table_name, table_type
    -> FROM information_schema.tables
    -> WHERE table_schema = 'bank' AND table_type = 'BASE TABLE'
    -> ORDER BY 1;
+--------------+------------+
| table_name   | table_type |
+--------------+------------+
| account      | BASE TABLE |
| branch       | BASE TABLE |
| business     | BASE TABLE |
| customer     | BASE TABLE |
| department   | BASE TABLE |
| employee     | BASE TABLE |
| individual   | BASE TABLE |
| officer      | BASE TABLE |
| product      | BASE TABLE |
| product_type | BASE TABLE |
| transaction  | BASE TABLE |
+--------------+------------+
11 rows in set (0.01 sec)
```

If you are only interested in information about views, you can query information_schema.views. Along with the view names, you can retrieve additional information, such as a flag that shows whether a view is updatable:

```
mysql> SELECT table_name, is_updatable
    -> FROM information_schema.views
    -> WHERE table_schema = 'bank'
    -> ORDER BY 1;
+----------------------+--------------+
| table_name           | is_updatable |
+----------------------+--------------+
| branch_activity_vw   | NO           |
| business_customer_vw | YES          |
| customer_vw          | YES          |
| employee_vw          | YES          |
```

```
| nh_customer_vw      | YES          |
+---------------------+--------------+
5 rows in set (1.83 sec)
```

Additionally, you can retrieve the view's underlying query using the `view_definition` column, as long as the query is small enough (4,000 characters or fewer for MySQL).

Column information for both tables and views is available via the `columns` view. The following query shows column information for the `account` table:

```
mysql> SELECT column_name, data_type, character_maximum_length char_max_len,
    ->   numeric_precision num_prcsn, numeric_scale num_scale
    -> FROM information_schema.columns
    -> WHERE table_schema = 'bank' AND table_name = 'account'
    -> ORDER BY ordinal_position;
+--------------------+-----------+--------------+-----------+-----------+
| column_name        | data_type | char_max_len | num_prcsn | num_scale |
+--------------------+-----------+--------------+-----------+-----------+
| account_id         | int       |         NULL |        10 |         0 |
| product_cd         | varchar   |           10 |      NULL |      NULL |
| cust_id            | int       |         NULL |        10 |         0 |
| open_date          | date      |         NULL |      NULL |      NULL |
| close_date         | date      |         NULL |      NULL |      NULL |
| last_activity_date | date      |         NULL |      NULL |      NULL |
| status             | enum      |            6 |      NULL |      NULL |
| open_branch_id     | smallint  |         NULL |         5 |         0 |
| open_emp_id        | smallint  |         NULL |         5 |         0 |
| avail_balance      | float     |         NULL |        10 |         2 |
| pending_balance    | float     |         NULL |        10 |         2 |
+--------------------+-----------+--------------+-----------+-----------+
11 rows in set (0.02 sec)
```

The `ordinal_position` column is included merely as a means to retrieve the columns in the order in which they were added to the table.

You can retrieve information about a table's indexes via the `information_schema.sta tistics` view as demonstrated by the following query, which retrieves information for the indexes built on the `account` table:

```
mysql> SELECT index_name, non_unique, seq_in_index, column_name
    -> FROM information_schema.statistics
    -> WHERE table_schema = 'bank' AND table_name = 'account'
    -> ORDER BY 1, 3;
+----------------+------------+--------------+----------------+
| index_name     | non_unique | seq_in_index | column_name    |
+----------------+------------+--------------+----------------+
| acc_bal_idx    |          1 |            1 | cust_id        |
| acc_bal_idx    |          1 |            2 | avail_balance  |
| fk_a_branch_id |          1 |            1 | open_branch_id |
| fk_a_emp_id    |          1 |            1 | open_emp_id    |
| fk_product_cd  |          1 |            1 | product_cd     |
| PRIMARY        |          0 |            1 | account_id     |
+----------------+------------+--------------+----------------+
6 rows in set (0.09 sec)
```

The `account` table has a total of five indexes, one of which has two columns (`acc_bal_idx`) and one of which is a unique index (`PRIMARY`).

You can retrieve the different types of constraints (foreign key, primary key, unique) that have been created via the `information_schema.table_constraints` view. Here's a query that retrieves all of the constraints in the `bank` schema:

```
mysql> SELECT constraint_name, table_name, constraint_type
    -> FROM information_schema.table_constraints
    -> WHERE table_schema = 'bank'
    -> ORDER BY 3,1;
+--------------------+--------------+-----------------+
| constraint_name    | table_name   | constraint_type |
+--------------------+--------------+-----------------+
| fk_a_branch_id     | account      | FOREIGN KEY     |
| fk_a_cust_id       | account      | FOREIGN KEY     |
| fk_a_emp_id        | account      | FOREIGN KEY     |
| fk_b_cust_id       | business     | FOREIGN KEY     |
| fk_dept_id         | employee     | FOREIGN KEY     |
| fk_exec_branch_id  | transaction  | FOREIGN KEY     |
| fk_e_branch_id     | employee     | FOREIGN KEY     |
| fk_e_emp_id        | employee     | FOREIGN KEY     |
| fk_i_cust_id       | individual   | FOREIGN KEY     |
| fk_o_cust_id       | officer      | FOREIGN KEY     |
| fk_product_cd      | account      | FOREIGN KEY     |
| fk_product_type_cd | product      | FOREIGN KEY     |
| fk_teller_emp_id   | transaction  | FOREIGN KEY     |
| fk_t_account_id    | transaction  | FOREIGN KEY     |
| PRIMARY            | branch       | PRIMARY KEY     |
| PRIMARY            | account      | PRIMARY KEY     |
| PRIMARY            | product      | PRIMARY KEY     |
| PRIMARY            | department   | PRIMARY KEY     |
| PRIMARY            | customer     | PRIMARY KEY     |
| PRIMARY            | transaction  | PRIMARY KEY     |
| PRIMARY            | officer      | PRIMARY KEY     |
| PRIMARY            | product_type | PRIMARY KEY     |
| PRIMARY            | employee     | PRIMARY KEY     |
| PRIMARY            | business     | PRIMARY KEY     |
| PRIMARY            | individual   | PRIMARY KEY     |
| dept_name_idx      | department   | UNIQUE          |
+--------------------+--------------+-----------------+
26 rows in set (2.28 sec)
```

Table 15-1 shows the entire set of `information_schema` views that are available in MySQL version 6.0.

Table 15-1. Information_schema views

View name	Provides information about…
Schemata	Databases
Tables	Tables and views
Columns	Columns of tables and views
Statistics	Indexes

View name	Provides information about...
User_Privileges	Who has privileges on which schema objects
Schema_Privileges	Who has privileges on which databases
Table_Privileges	Who has privileges on which tables
Column_Privileges	Who has privileges on which columns of which tables
Character_Sets	What character sets are available
Collations	What collations are available for which character sets
Collation_Character_Set_Applicability	Which character sets are available for which collation
Table_Constraints	The unique, foreign key, and primary key constraints
Key_Column_Usage	The constraints associated with each key column
Routines	Stored routines (procedures and functions)
Views	Views
Triggers	Table triggers
Plugins	Server plug-ins
Engines	Available storage engines
Partitions	Table partitions
Events	Scheduled events
Process_List	Running processes
Referential_Constraints	Foreign keys
Global_Status	Server status information
Session_Status	Session status information
Global_Variables	Server status variables
Session_Variables	Session status variables
Parameters	Stored procedure and function parameters
Profiling	User profiling information

While some of these views, such as engines, events, and plugins, are specific to MySQL, many of these views are available in SQL Server as well. If you are using Oracle Database, please consult the online Oracle Database Reference Guide (*http://www.oracle.com/pls/db111/portal.all_books*) for information about the user_, all_, and dba_ views.

Working with Metadata

As I mentioned earlier, having the ability to retrieve information about your schema objects via SQL queries opens up some interesting possibilities. This section shows several ways in which you can make use of metadata in your applications.

Schema Generation Scripts

While some project teams include a full-time database designer who oversees the design and implementation of the database, many projects take the "design-by-committee" approach, allowing multiple people to create database objects. After several weeks or months of development, you may need to generate a script that will create the various tables, indexes, views, and so on that the team has deployed. Although a variety of tools and utilities will generate these types of scripts for you, you can also query the `information_schema` views and generate the script yourself.

As an example, let's build a script that will create the `bank.customer` table. Here's the command used to build the table, which I extracted from the script used to build the example database:

```
create table customer
 (cust_id integer unsigned not null auto_increment,
  fed_id varchar(12) not null,
  cust_type_cd enum('I','B') not null,
  address varchar(30),
  city varchar(20),
  state varchar(20),
  postal_code varchar(10),
  constraint pk_customer primary key (cust_id)
 );
```

Although it would certainly be easier to generate the script with the use of a procedural language (e.g., Transact-SQL or Java), since this is a book about SQL I'm going to write a single query that will generate the `create table` statement. The first step is to query the `information_schema.columns` table to retrieve information about the columns in the table:

```
mysql> SELECT 'CREATE TABLE customer (' create_table_statement
    -> UNION ALL
    -> SELECT cols.txt
    -> FROM
    -> (SELECT concat('  ',column_name, ' ', column_type,
    ->     CASE
    ->       WHEN is_nullable = 'NO' THEN ' not null'
    ->       ELSE ''
    ->     END,
    ->     CASE
    ->       WHEN extra IS NOT NULL THEN concat(' ', extra)
    ->       ELSE ''
    ->     END,
    ->     ',') txt
    ->   FROM information_schema.columns
    ->   WHERE table_schema = 'bank' AND table_name = 'customer'
    ->   ORDER BY ordinal_position
    -> ) cols
    -> UNION ALL
    -> SELECT ')';
+----------------------------------------------------+
| create_table_statement                             |
```

```
+-------------------------------------------------------+
| CREATE TABLE customer (                               |
|   cust_id int(10) unsigned not null auto_increment,   |
|   fed_id varchar(12) not null ,                       |
|   cust_type_cd enum('I','B') not null ,               |
|   address varchar(30) ,                               |
|   city varchar(20) ,                                  |
|   state varchar(20) ,                                 |
|   postal_code varchar(10) ,                           |
| )                                                     |
+-------------------------------------------------------+
9 rows in set (0.04 sec)
```

Well, that got us pretty close; we just need to add queries against the table_constraints and key_column_usage views to retrieve information about the primary key constraint:

```
mysql> SELECT 'CREATE TABLE customer (' create_table_statement
    -> UNION ALL
    -> SELECT cols.txt
    -> FROM
    ->   (SELECT concat('  ',column_name, ' ', column_type,
    ->      CASE
    ->        WHEN is_nullable = 'NO' THEN ' not null'
    ->        ELSE ''
    ->      END,
    ->      CASE
    ->        WHEN extra IS NOT NULL THEN concat(' ', extra)
    ->        ELSE ''
    ->      END,
    ->      ',') txt
    ->    FROM information_schema.columns
    ->    WHERE table_schema = 'bank' AND table_name = 'customer'
    ->    ORDER BY ordinal_position
    ->   ) cols
    -> UNION ALL
    -> SELECT concat('  constraint primary key (')
    -> FROM information_schema.table_constraints
    -> WHERE table_schema = 'bank' AND table_name = 'customer'
    ->   AND constraint_type = 'PRIMARY KEY'
    -> UNION ALL
    -> SELECT cols.txt
    -> FROM
    ->   (SELECT concat(CASE WHEN ordinal_position > 1 THEN '   ,'
    ->      ELSE '    ' END, column_name) txt
    ->    FROM information_schema.key_column_usage
    ->    WHERE table_schema = 'bank' AND table_name = 'customer'
    ->      AND constraint_name = 'PRIMARY'
    ->    ORDER BY ordinal_position
    ->   ) cols
    -> UNION ALL
    -> SELECT '  )'
    -> UNION ALL
    -> SELECT ')';
+-------------------------------------------------------+
```

```
| create_table_statement                             |
+-----------------------------------------------------+
| CREATE TABLE customer (                             |
|   cust_id int(10) unsigned not null auto_increment, |
|   fed_id varchar(12) not null ,                     |
|   cust_type_cd enum('I','B') not null ,             |
|   address varchar(30) ,                             |
|   city varchar(20) ,                                |
|   state varchar(20) ,                               |
|   postal_code varchar(10) ,                         |
|   constraint primary key (                          |
|     cust_id                                         |
|   )                                                 |
| )                                                   |
+-----------------------------------------------------+
12 rows in set (0.02 sec)
```

To see whether the statement is properly formed, I'll paste the query output into the mysql tool (I've changed the table name to customer2 so that it won't step on our other table):

```
mysql> CREATE TABLE customer2 (
    ->    cust_id int(10) unsigned not null auto_increment,
    ->    fed_id varchar(12) not null ,
    ->    cust_type_cd enum('I','B') not null ,
    ->    address varchar(30) ,
    ->    city varchar(20) ,
    ->    state varchar(20) ,
    ->    postal_code varchar(10) ,
    ->    constraint primary key (
    ->       cust_id
    ->    )
    -> );
Query OK, 0 rows affected (0.14 sec)
```

The statement executed without errors, and there is now a customer2 table in the bank database. In order for the query to generate a well-formed create table statement for *any* table, more work is required (such as handling indexes and foreign key constraints), but I'll leave that as an exercise.

Deployment Verification

Many organizations allow for database maintenance windows, wherein existing database objects may be administered (such as adding/dropping partitions) and new schema objects and code can be deployed. After the deployment scripts have been run, it's a good idea to run a verification script to ensure that the new schema objects are in place with the appropriate columns, indexes, primary keys, and so forth. Here's a query that returns the number of columns, number of indexes, and number of primary key constraints (0 or 1) for each table in the bank schema:

```
mysql> SELECT tbl.table_name,
    ->  (SELECT count(*) FROM information_schema.columns clm
    ->   WHERE clm.table_schema = tbl.table_schema
    ->     AND clm.table_name = tbl.table_name) num_columns,
    ->  (SELECT count(*) FROM information_schema.statistics sta
    ->   WHERE sta.table_schema = tbl.table_schema
    ->     AND sta.table_name = tbl.table_name) num_indexes,
    ->  (SELECT count(*) FROM information_schema.table_constraints tc
    ->   WHERE tc.table_schema = tbl.table_schema
    ->     AND tc.table_name = tbl.table_name
    ->     AND tc.constraint_type = 'PRIMARY KEY') num_primary_keys
    -> FROM information_schema.tables tbl
    -> WHERE tbl.table_schema = 'bank' AND tbl.table_type = 'BASE TABLE'
    -> ORDER BY 1;
+--------------+-------------+-------------+-------------------+
| table_name   | num_columns | num_indexes | num_primary_keys  |
+--------------+-------------+-------------+-------------------+
| account      |          11 |           6 |                 1 |
| branch       |           6 |           1 |                 1 |
| business     |           4 |           1 |                 1 |
| customer     |           7 |           1 |                 1 |
| department   |           2 |           2 |                 1 |
| employee     |           9 |           4 |                 1 |
| individual   |           4 |           1 |                 1 |
| officer      |           7 |           2 |                 1 |
| product      |           5 |           2 |                 1 |
| product_type |           2 |           1 |                 1 |
| transaction  |           8 |           4 |                 1 |
+--------------+-------------+-------------+-------------------+
11 rows in set (13.83 sec)
```

You could execute this statement before and after the deployment and then verify any differences between the two sets of results before declaring the deployment a success.

Dynamic SQL Generation

Some languages, such as Oracle's PL/SQL and Microsoft's Transact-SQL, are supersets of the SQL language, meaning that they include SQL statements in their grammar along with the usual procedural constructs, such as "if-then-else" and "while." Other languages, such as Java, include the ability to interface with a relational database, but do not include SQL statements in the grammar, meaning that all SQL statements must be contained within strings.

Therefore, most relational database servers, including SQL Server, Oracle Database, and MySQL, allow SQL statements to be submitted to the server as strings. Submitting strings to a database engine rather than utilizing its SQL interface is generally known as *dynamic SQL execution*. Oracle's PL/SQL language, for example, includes an execute immediate command, which you can use to submit a string for execution, while SQL Server includes a system stored procedure called sp_executesql for executing SQL statements dynamically.

MySQL provides the statements `prepare`, `execute`, and `deallocate` to allow for dynamic SQL execution. Here's a simple example:

```
mysql> SET @qry = 'SELECT cust_id, cust_type_cd, fed_id FROM customer';
Query OK, 0 rows affected (0.07 sec)

mysql> PREPARE dynsql1 FROM @qry;
Query OK, 0 rows affected (0.04 sec)
Statement prepared

mysql> EXECUTE dynsql1;
+---------+--------------+-------------+
| cust_id | cust_type_cd | fed_id      |
+---------+--------------+-------------+
|       1 | I            | 111-11-1111 |
|       2 | I            | 222-22-2222 |
|       3 | I            | 333-33-3333 |
|       4 | I            | 444-44-4444 |
|       5 | I            | 555-55-5555 |
|       6 | I            | 666-66-6666 |
|       7 | I            | 777-77-7777 |
|       8 | I            | 888-88-8888 |
|       9 | I            | 999-99-9999 |
|      10 | B            | 04-1111111  |
|      11 | B            | 04-2222222  |
|      12 | B            | 04-3333333  |
|      13 | B            | 04-4444444  |
|      99 | I            | 04-9999999  |
+---------+--------------+-------------+
14 rows in set (0.27 sec)

mysql> DEALLOCATE PREPARE dynsql1;
Query OK, 0 rows affected (0.00 sec)
```

The `set` statement simply assigns a string to the `qry` variable, which is then submitted to the database engine (for parsing, security checking, and optimization) using the `prepare` statement. After executing the statement by calling `execute`, the statement must be closed using `deallocate prepare`, which frees any database resources (e.g., cursors) that have been utilized during execution.

The next example shows how you could execute a query that includes placeholders so that conditions can be specified at runtime:

```
mysql> SET @qry = 'SELECT product_cd, name, product_type_cd, date_offered, date_
retired FROM product WHERE product_cd = ?';
Query OK, 0 rows affected (0.00 sec)

mysql> PREPARE dynsql2 FROM @qry;
Query OK, 0 rows affected (0.00 sec)
Statement prepared

mysql> SET @prodcd = 'CHK';
Query OK, 0 rows affected (0.00 sec)

mysql> EXECUTE dynsql2 USING @prodcd;
```

```
+------------+-----------------+-----------------+--------------+--------------+
| product_cd | name            | product_type_cd | date_offered | date_retired |
+------------+-----------------+-----------------+--------------+--------------+
| CHK        | checking account | ACCOUNT        | 2004-01-01   | NULL         |
+------------+-----------------+-----------------+--------------+--------------+
1 row in set (0.01 sec)

mysql> SET @prodcd = 'SAV';
Query OK, 0 rows affected (0.00 sec)

mysql> EXECUTE dynsql2 USING @prodcd;
+------------+-----------------+-----------------+--------------+--------------+
| product_cd | name            | product_type_cd | date_offered | date_retired |
+------------+-----------------+-----------------+--------------+--------------+
| SAV        | savings account | ACCOUNT         | 2004-01-01   | NULL         |
+------------+-----------------+-----------------+--------------+--------------+
1 row in set (0.00 sec)

mysql> DEALLOCATE PREPARE dynsql2;
Query OK, 0 rows affected (0.00 sec)
```

In this sequence, the query contains a placeholder (the ? at the end of the statement)
so that the product code can be submitted at runtime. The statement is prepared once
and then executed twice, once for product code 'CHK' and again for product code
'SAV', after which the statement is closed.

What, you may wonder, does this have to do with metadata? Well, if you are going to
use dynamic SQL to query a table, why not build the query string using metadata rather
than hardcoding the table definition? The following example generates the same dynamic SQL string as the previous example, but it retrieves the column names from the
information_schema.columns view:

```
mysql> SELECT concat('SELECT ',
    ->   concat_ws(',', cols.col1, cols.col2, cols.col3, cols.col4,
    ->     cols.col5, cols.col6, cols.col7, cols.col8, cols.col9),
    ->   ' FROM product WHERE product_cd = ?')
    -> INTO @qry
    -> FROM
    ->   (SELECT
    ->     max(CASE WHEN ordinal_position = 1 THEN column_name
    ->       ELSE NULL END) col1,
    ->     max(CASE WHEN ordinal_position = 2 THEN column_name
    ->       ELSE NULL END) col2,
    ->     max(CASE WHEN ordinal_position = 3 THEN column_name
    ->       ELSE NULL END) col3,
    ->     max(CASE WHEN ordinal_position = 4 THEN column_name
    ->       ELSE NULL END) col4,
    ->     max(CASE WHEN ordinal_position = 5 THEN column_name
    ->       ELSE NULL END) col5,
    ->     max(CASE WHEN ordinal_position = 6 THEN column_name
    ->       ELSE NULL END) col6,
    ->     max(CASE WHEN ordinal_position = 7 THEN column_name
    ->       ELSE NULL END) col7,
    ->     max(CASE WHEN ordinal_position = 8 THEN column_name
```

```
   ->        ELSE NULL END) col8,
   ->      max(CASE WHEN ordinal_position = 9 THEN column_name
   ->        ELSE NULL END) col9
   ->   FROM information_schema.columns
   ->   WHERE table_schema = 'bank' AND table_name = 'product'
   ->   GROUP BY table_name
   -> ) cols;
Query OK, 1 row affected (0.02 sec)

mysql> SELECT @qry;
+--------------------------------------------------------------------------
----------------------+
| @qry
                      |
+--------------------------------------------------------------------------
----------------------+
| SELECT product_cd,name,product_type_cd,date_offered,date_retired FROM product
WHERE product_cd = ? |
+--------------------------------------------------------------------------
----------------------+
1 row in set (0.00 sec)

mysql> PREPARE dynsql3 FROM @qry;
Query OK, 0 rows affected (0.01 sec)
Statement prepared

mysql> SET @prodcd = 'MM';
Query OK, 0 rows affected (0.00 sec)

mysql> EXECUTE dynsql3 USING @prodcd;
+------------+----------------------+-----------------+--------------+--------------+
| product_cd | name                 | product_type_cd | date_offered | date_retired |
+------------+----------------------+-----------------+--------------+--------------+
| MM         | money market account | ACCOUNT         | 2004-01-01   | NULL         |
+------------+----------------------+-----------------+--------------+--------------+
1 row in set (0.00 sec)

mysql> DEALLOCATE PREPARE dynsql3;
Query OK, 0 rows affected (0.00 sec)
```

The query pivots the first nine columns in the product table, builds a query string using the concat and concat_ws functions, and assigns the string to the qry variable. The query string is then executed as before.

> Generally, it would be better to generate the query using a procedural language that includes looping constructs, such as Java, PL/SQL, Transact-SQL, or MySQL's Stored Procedure Language. However, I wanted to demonstrate a pure SQL example, so I had to limit the number of columns retrieved to some reasonable number, which in this example is nine.

Test Your Knowledge

The following exercises are designed to test your understanding of metadata. When you're finished, please see Appendix C for the solutions.

Exercise 15-1

Write a query that lists all of the indexes in the bank schema. Include the table names.

Exercise 15-2

Write a query that generates output that can be used to create all of the indexes on the bank.employee table. Output should be of the form:

```
"ALTER TABLE <table_name> ADD INDEX <index_name> (<column_list>)"
```

ER Diagram for Example Database

Figure A-1 is an entity-relationship (ER) diagram for the example database used in this book. As the name suggests, the diagram depicts the entities, or tables, in the database along with the foreign-key relationships between the tables. Here are a few tips to help you understand the notation:

- Each rectangle represents a table, with the table name above the upper-left corner of the rectangle. The primary-key column(s) are listed first and are separated from nonkey columns by a line. Nonkey columns are listed below the line, and foreign key columns are marked with "(FK)."

- Lines between tables represent foreign key relationships. The markings at either end of the lines represents the allowable quantity, which can be zero (0), one (1), or many (≤). For example, if you look at the relationship between the account and product tables, you would say that an account must belong to exactly one product, but a product may have zero, one, or many accounts.

For more information on entity-relationship modeling, please see *http://en.wikipedia .org/wiki/Entity-relationship_model*.

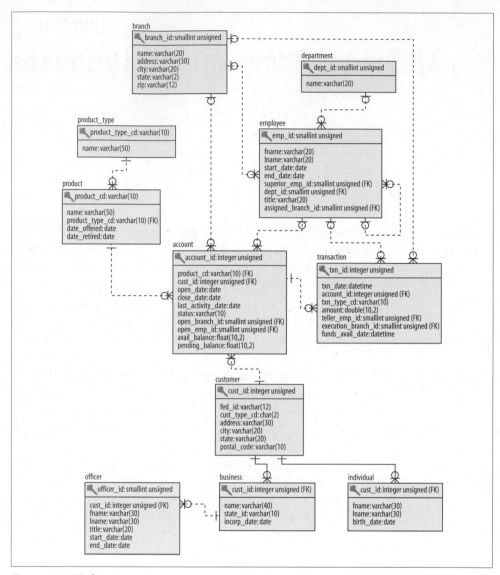

Figure A-1. ER diagram

MySQL Extensions to the SQL Language

Since this book uses the MySQL server for all the examples, I thought it would be useful for readers who are planning to continue using MySQL to include an appendix on MySQL's extensions to the SQL language. This appendix explores some of MySQL's extensions to the `select`, `insert`, `update`, and `delete` statements that can be very useful in certain situations.

Extensions to the select Statement

MySQL's implementation of the `select` statement includes two additional clauses, which are discussed in the following subsections.

The limit Clause

In some situations, you may not be interested in *all* of the rows returned by a query. For example, you might construct a query that returns all of the bank tellers along with the number of accounts opened by each teller. If your reason for executing the query is to determine the top three tellers so that they can receive an award from the bank, then you don't necessarily need to know who came in fourth, fifth, and so on. To help with these types of situations, MySQL's `select` statement includes the `limit` clause, which allows you to restrict the number of rows returned by a query.

To demonstrate the utility of the `limit` clause, I will begin by constructing a query to show the number of accounts opened by each bank teller:

```
mysql> SELECT open_emp_id, COUNT(*) how_many
    -> FROM account
    -> GROUP BY open_emp_id;
+-------------+----------+
| open_emp_id | how_many |
+-------------+----------+
|           1 |        8 |
|          10 |        7 |
|          13 |        3 |
|          16 |        6 |
```

```
+-------------+----------+
4 rows in set (0.31 sec)
```

The results show that four different tellers opened accounts; if you want to limit the result set to only three records, you can add a `limit` clause specifying that only three records be returned:

```
mysql> SELECT open_emp_id, COUNT(*) how_many
    -> FROM account
    -> GROUP BY open_emp_id
    -> LIMIT 3;
+-------------+----------+
| open_emp_id | how_many |
+-------------+----------+
|           1 |        8 |
|          10 |        7 |
|          13 |        3 |
+-------------+----------+
3 rows in set (0.06 sec)
```

Thanks to the `limit` clause (the fourth line of the query), the result set now includes exactly three records, and the fourth teller (employee ID 16) has been discarded from the result set.

Combining the limit clause with the order by clause

While the previous query returns three records, there's one small problem; you haven't described *which* three of the four records you are interested in. If you are looking for three *specific* records, such as the three tellers who opened the most accounts, you will need to use the `limit` clause in concert with an **order** by clause, as in:

```
mysql> SELECT open_emp_id, COUNT(*) how_many
    -> FROM account
    -> GROUP BY open_emp_id
    -> ORDER BY how_many DESC
    -> LIMIT 3;
+-------------+----------+
| open_emp_id | how_many |
+-------------+----------+
|           1 |        8 |
|          10 |        7 |
|          16 |        6 |
+-------------+----------+
3 rows in set (0.03 sec)
```

The difference between this query and the previous query is that the `limit` clause is now being applied to an ordered set, resulting in the three tellers with the most opened accounts being included in the final result set. Unless you are interested in seeing only an arbitrary sample of records, you will generally want to use an **order** by clause along with a `limit` clause.

The `limit` clause is applied after all filtering, grouping, and ordering have occurred, so it will never change the outcome of your `select` statement other than restricting the number of records returned by the statement.

The limit clause's optional second parameter

Instead of finding the top three tellers, let's say your goal is to identify all but the top two tellers (instead of giving awards to top performers, the bank will be sending some of the less-productive tellers to assertiveness training). For these types of situations, the `limit` clause allows for an optional second parameter; when two parameters are used, the first designates at which record to begin adding records to the final result set, and the second designates how many records to include. When specifying a record by number, remember that MySQL designates the first record as record 0. Therefore, if your goal is to find the third-best performer, you can do the following:

```
mysql> SELECT open_emp_id, COUNT(*) how_many
    -> FROM account
    -> GROUP BY open_emp_id
    -> ORDER BY how_many DESC
    -> LIMIT 2, 1;
+-------------+----------+
| open_emp_id | how_many |
+-------------+----------+
|          16 |        6 |
+-------------+----------+
1 row in set (0.00 sec)
```

In this example, the zeroth and first records are discarded, and records are included starting at the second record. Since the second parameter in the `limit` clause is 1, only a single record is included.

If you want to start at the second position and include *all* the remaining records, you can make the second argument to the `limit` clause large enough to guarantee that all remaining records are included. If you do not know how many tellers opened new accounts, therefore, you might do something like the following to find all but the top two performers:

```
mysql> SELECT open_emp_id, COUNT(*) how_many
    -> FROM account
    -> GROUP BY open_emp_id
    -> ORDER BY how_many DESC
    -> LIMIT 2, 999999999;
+-------------+----------+
| open_emp_id | how_many |
+-------------+----------+
|          16 |        6 |
|          13 |        3 |
+-------------+----------+
2 rows in set (0.00 sec)
```

In this version of the query, the zeroth and first records are discarded, and up to 999,999,999 records are included starting at the second record (in this case, there are only two more, but it's better to go a bit overboard rather than taking a chance on excluding valid records from your final result set because you underestimated).

Ranking queries

When used in conjunction with an `order by` clause, queries that include a `limit` clause can be called *ranking queries* because they allow you to rank your data. While I have demonstrated how to rank bank tellers by the number of opened accounts, ranking queries are used to answer many different types of business questions, such as:

- Who are our top five salespeople for 2005?
- Who has the third-most home runs in the history of baseball?
- Other than *The Holy Bible* and *Quotations from Chairman Mao*, what are the next 98 best-selling books of all time?
- What are our two worst-selling flavors of ice cream?

So far, I have shown how to find the top three tellers, the third-best teller, and all but the top two tellers. If I want to do something analogous to the fourth example (i.e., find the *worst* performers), I need only reverse the sort order so that the results proceed from lowest number of accounts opened to highest number of accounts opened, as in:

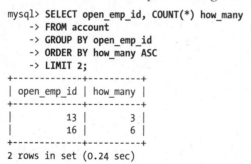

```
mysql> SELECT open_emp_id, COUNT(*) how_many
    -> FROM account
    -> GROUP BY open_emp_id
    -> ORDER BY how_many ASC
    -> LIMIT 2;
+-------------+----------+
| open_emp_id | how_many |
+-------------+----------+
|          13 |        3 |
|          16 |        6 |
+-------------+----------+
2 rows in set (0.24 sec)
```

By simply changing the sort order (from `ORDER BY how_many DESC` to `ORDER BY how_many ASC`), the query now returns the two worst-performing tellers. Therefore, by using a `limit` clause with either an ascending or descending sort order, you can produce ranking queries to answer most types of business questions.

The into outfile Clause

If you want the output from your query to be written to a file, you could highlight the query results, copy them to the buffer, and paste them into your favorite editor. However, if the query's result set is sufficiently large, or if the query is being executed from within a script, you will need a way to write the results to a file without your intervention. To aid in such situations, MySQL includes the `into outfile` clause to allow you to provide the name of a file into which the results will be written. Here's an example that writes the query results to a file in my *c:\temp* directory:

```
mysql> SELECT emp_id, fname, lname, start_date
    -> INTO OUTFILE 'C:\\TEMP\\emp_list.txt'
    -> FROM employee; Query OK, 18 rows affected (0.20 sec)
```

 If you remember from Chapter 7, the backslash is used to escape another character within a string. If you're a Windows user, therefore, you will need to enter two backslashes in a row when building pathnames.

Rather than showing the query results on the screen, the result set has been written to the *emp_list.txt* file, which looks as follows:

```
1     Michael     Smith     2001-06-22
2     Susan       Barker    2002-09-12
3     Robert      Tyler     2000-02-09
4     Susan       Hawthorne 2002-04-24
...
16    Theresa     Markham   2001-03-15
17    Beth        Fowler    2002-06-29
18    Rick        Tulman    2002-12-12
```

The default format uses tabs ('\t') between columns and newlines ('\n') after each record. If you want more control over the format of the data, several additional sub-clauses are available with the `into outfile` clause. For example, if you want the data to be in what is referred to as *pipe-delimited format*, you can use the `fields` subclause to ask that the '|' character be placed between each column, as in:

```
mysql> SELECT emp_id, fname, lname, start_date
    -> INTO OUTFILE 'C:\\TEMP\\emp_list_delim.txt'
    ->    FIELDS TERMINATED BY '|'
    -> FROM employee; Query OK, 18 rows affected (0.02 sec)
```

 MySQL does not allow you to overwrite an existing file when using `into outfile`, so you will need to remove an existing file first if you run the same query more than once.

The contents of the *emp_list_delim.txt* file look as follows:

```
1|Michael|Smith|2001-06-22
2|Susan|Barker|2002-09-12
3|Robert|Tyler|2000-02-09
4|Susan|Hawthorne|2002-04-24
...
16|Theresa|Markham|2001-03-15
17|Beth|Fowler|2002-06-29
18|Rick|Tulman|2002-12-12
```

Along with pipe-delimited format, you may need your data in *comma-delimited format*, in which case you would use `fields terminated by ','`. If the data being written to a file includes strings, however, using commas as field separators can prove problematic, as commas are much more likely to appear within strings than the pipe character. Consider the following query, which writes a number and two strings delimited by commas to the *comma1.txt* file:

```
mysql> SELECT data.num, data.str1, data.str2
    -> INTO OUTFILE 'C:\\TEMP\\comma1.txt'
    ->   FIELDS TERMINATED BY ','
    -> FROM
    -> (SELECT 1 num, 'This string has no commas' str1,
    ->     'This string, however, has two commas' str2) data;
Query OK, 1 row affected (0.04 sec)
```

Since the third column in the output file (`str2`) is a string containing commas, you might think that an application attempting to read the *comma1.txt* file will encounter problems when parsing each line into columns, but the MySQL server has made provisions for such situations. Here are the contents of *comma1.txt*:

```
1,This string has no commas,This string\, however\, has two commas
```

As you can see, the commas within the third column have been escaped by putting a backslash before the two commas embedded in the `str2` column. If you run the same query but use pipe-delimited format, the commas will *not* be escaped, since they don't need to be. If you want to use a different escape character, such as using another comma, you can use the `fields escaped by` subclause to specify the escape character to use for your output file.

Along with specifying column separators, you can also specify the character used to separate the different records in your datafile. If you would like each record in the output file to be separated by something other than the newline character, you can use the `lines` subclause, as in:

```
mysql> SELECT emp_id, fname, lname, start_date
    -> INTO OUTFILE 'C:\\TEMP\\emp_list_atsign.txt'
    ->   FIELDS TERMINATED BY '|'
    ->   LINES TERMINATED BY '@'
    -> FROM employee;
Query OK, 18 rows affected (0.03 sec)
```

Because I am not using a newline character between records, the *emp_list_atsign.txt* file looks like a single long line of text when viewed, with each record separated by the '@' character:

```
1|Michael|Smith|2001-06-22@2|Susan|Barker|2002-09-12@3|Robert|Tyler|2000-02-
09@4|Susan|Hawthorne|2002-04-24@5|John|Gooding|2003-11-14@6|Helen|Fleming|2004-03-
17@7|Chris|Tucker|2004-09-15@8|Sarah|Parker|2002-12-02@9|Jane|Grossman|2002-05-
03@10|Paula|Roberts|2002-07-27@11|Thomas|Ziegler|2000-10-23@12|Samantha|Jameson|2003-
01-08@13|John|Blake|2000-05-11@14|Cindy|Mason|2002-08-09@15|Frank|Portman|2003-04-
01@16|Theresa|Markham|2001-03-15@17|Beth|Fowler|2002-06-29@18|Rick|Tulman|2002-12-12@
```

If you need to generate a datafile to be loaded into a spreadsheet application or sent within or outside your organization, the `into outfile` clause should provide enough flexibility for whatever file format you need.

Combination Insert/Update Statements

Let's say that you have been asked to create a table to capture information about which of the bank's branches are visited by which customers. The table needs to contain the customer's ID, the branch's ID, and a `datetime` column indicating the last time the customer visited the branch. Rows are added to the table whenever a customer visits a certain branch, but if the customer has already visited the branch, then the existing row should simply have its `datetime` column updated. Here's the table definition:

```
CREATE TABLE branch_usage
 (branch_id SMALLINT UNSIGNED NOT NULL,
  cust_id INTEGER UNSIGNED NOT NULL,
  last_visited_on DATETIME,
  CONSTRAINT pk_branch_usage PRIMARY KEY (branch_id, cust_id)
 );
```

Along with the three column definitions, the `branch_usage` table defines a primary key constraint on the `branch_id` and `cust_id` columns. Therefore, the server will reject any row added to the table whose branch/customer pair already exists in the table.

Let's say that, after the table is in place, customer ID 5 visits the main branch (branch ID 1) three times in the first week. After the first visit, you can insert a record into the `branch_usage` table, since no record exists yet for customer ID 5 and branch ID 1:

```
mysql> INSERT INTO branch_usage (branch_id, cust_id, last_visited_on)
    -> VALUES (1, 5, CURRENT_TIMESTAMP());
Query OK, 1 row affected (0.02 sec)
```

The next time the customer visits the same branch, however, you will need to *update* the existing record rather than inserting a new record; otherwise, you will receive the following error:

```
ERROR 1062 (23000): Duplicate entry '1-5' for key 1
```

To avoid this error, you can query the `branch_usage` table to see whether a given customer/branch pair exists and then either insert a record if no record is found or update the existing row if it already exists. To save you the trouble, however, the MySQL designers have extended the `insert` statement to allow you to specify that one or more columns be modified if an `insert` statement fails due to a duplicate key. The following statement instructs the server to modify the `last_visited_on` column if the given customer and branch already exist in the `branch_usage` table:

```
mysql> INSERT INTO branch_usage (branch_id, cust_id, last_visited_on)
    -> VALUES (1, 5, CURRENT_TIMESTAMP())
    -> ON DUPLICATE KEY UPDATE last_visited_on = CURRENT_TIMESTAMP();
Query OK, 2 rows affected (0.02 sec)
```

The on duplicate key clause allows this same statement to be executed every time customer ID 5 conducts business in branch ID 1. If run 100 times, the first execution results in a single row being added to the table, and the next 99 executions result in the last_visited_on column being changed to the current time. This type of operation is often referred to as an *upsert*, since it is a combination of an update and an insert statement.

Replacing the replace Command

Prior to version 4.1 of the MySQL server, upsert operations were performed using the replace command, which is a proprietary statement that first deletes an existing row if the primary key value already exists in the table before inserting a row. If you are using version 4.1 or later, you can choose between the replace command and the insert...on duplicate key command when performing upsert operations.

However, the replace command performs a delete operation when duplicate key values are encountered, which can cause a ripple effect if you are using the InnoDB storage engine and have foreign key constraints enabled. If the constraints have been created with the on delete cascade option, then rows in other tables may also be automatically deleted when the replace command deletes a row in the target table. For this reason, it is generally regarded as safer to use the on duplicate key clause of the insert statement rather than the older replace command.

Ordered Updates and Deletes

Earlier in the appendix, I showed you how to write queries using the limit clause in conjunction with an order by clause to generate rankings, such as the top three tellers in terms of accounts opened. MySQL also allows the limit and order by clauses to be used in both update and delete statements, thereby allowing you to modify or remove specific rows in a table based on a ranking. For example, imagine that you are asked to remove records from a table used to track customer logins to the bank's online banking system. The table, which tracks the customer ID and date/time of login, looks as follows:

```
CREATE TABLE login_history
 (cust_id INTEGER UNSIGNED NOT NULL,
  login_date DATETIME,
  CONSTRAINT pk_login_history PRIMARY KEY (cust_id, login_date)
 );
```

The following statement populates the login_history table with some data by generating a cross join between the account and customer tables and using the account's open_date column as a basis for generating login dates:

```
mysql> INSERT INTO login_history (cust_id, login_date)
    -> SELECT c.cust_id,
    ->   ADDDATE(a.open_date, INTERVAL a.account_id * c.cust_id HOUR)
```

```
    -> FROM customer c CROSS JOIN account a;
Query OK, 312 rows affected (0.03 sec)
Records: 312  Duplicates: 0  Warnings: 0
```

The table is now populated with 312 rows of relatively random data. Your task is to look at the data in the login_history table once a month, generate a report for your manager showing who is using the online banking system, and then delete all but the 50 most-recent records from the table. One approach would be to write a query using order by and limit to find the 50th most recent login, such as:

```
mysql> SELECT login_date
    -> FROM login_history
    -> ORDER BY login_date DESC
    -> LIMIT 49,1;
+---------------------+
| login_date          |
+---------------------+
| 2004-07-02 09:00:00 |
+---------------------+
1 row in set (0.00 sec)
```

Armed with this information, you can then construct a delete statement that removes all rows whose login_date column is less than the date returned by the query:

```
mysql> DELETE FROM login_history
    -> WHERE login_date < '2004-07-02 09:00:00';
Query OK, 262 rows affected (0.02 sec)
```

The table now contains the 50 most-recent logins. Using MySQL's extensions, however, you can achieve the same result with a single delete statement using limit and order by clauses. After returning the original 312 rows to the login_history table, you can run the following:

```
mysql> DELETE FROM login_history
    -> ORDER BY login_date ASC
    -> LIMIT 262;
Query OK, 262 rows affected (0.05 sec)
```

With this statement, the rows are sorted by login_date in ascending order, and then the first 262 rows are deleted, leaving the 50 most recent rows.

> In this example, I had to know the number of rows in the table to construct the limit clause (312 original rows – 50 remaining rows = 262 deletions). It would be better if you could sort the rows in descending order and tell the server to skip the first 50 rows and then delete the remaining rows, as in:
>
> ```
> DELETE FROM login_history
> ORDER BY login_date DESC
> LIMIT 49, 9999999;
> ```
>
> However, MySQL does not allow the optional second parameter when using the limit clause in delete or update statements.

Along with deleting data, you can use the `limit` and `order by` clauses when modifying data as well. For example, if the bank decides to add $100 to each of the 10 oldest accounts to help retain loyal customers, you can do the following:

```
mysql> UPDATE account
    -> SET avail_balance = avail_balance + 100
    -> WHERE product_cd IN ('CHK', 'SAV', 'MM')
    -> ORDER BY open_date ASC
    -> LIMIT 10;
Query OK, 10 rows affected (0.06 sec)
Rows matched: 10  Changed: 10  Warnings: 0
```

This statement sorts accounts by the open date in ascending order and then modifies the first 10 records, which, in this case, are the 10 oldest accounts.

Multitable Updates and Deletes

In certain situations, you might need to modify or delete data from several different tables to perform a given task. If you discover that the bank's database contains a dummy customer left over from system testing, for example, you might need to remove data from the `account`, `customer`, and `individual` tables.

 For this section, I will create a set of clones for the `account`, `customer`, and `individual` tables, called `account2`, `customer2`, and `individual2`. I am doing so both to protect the sample data from being altered and to avoid any problems with foreign key constraints between the tables (more on this later in the section). Here are the `create table` statements used to generate the three clone tables:

```
CREATE TABLE individual2 AS
SELECT * FROM individual;
CREATE TABLE customer2 AS
SELECT * FROM customer;
CREATE TABLE account2 AS
SELECT * FROM account;
```

If the customer ID of the dummy customer is 1, you could generate three individual `delete` statements against each of the three tables, as in:

```
DELETE FROM account2
WHERE cust_id = 1;
DELETE FROM customer2
WHERE cust_id = 1;
DELETE FROM individual2
WHERE cust_id = 1;
```

Instead of writing individual `delete` statements, however, MySQL allows you to write a single *multitable* `delete` statement, which, in this case, looks as follows:

```
mysql> DELETE account2, customer2, individual2
    -> FROM account2 INNER JOIN customer2
    ->   ON account2.cust_id = customer2.cust_id
    ->   INNER JOIN individual2
    ->   ON customer2.cust_id = individual2.cust_id
    -> WHERE individual2.cust_id = 1;
Query OK, 5 rows affected (0.02 sec)
```

This statement removes a total of five rows, one from each of the individual2 and customer2 tables, and three from the account2 table (customer ID 1 has three accounts). The statement comprises three separate clauses:

delete
> Specifies the tables targeted for deletion.

from
> Specifies the tables used to identify the rows to be deleted. This clause is identical in form and function to the from clause in a select statement, and not all tables named herein need to be included in the delete clause.

where
> Contains filter conditions used to identify the rows to be deleted.

The multitable delete statement looks a lot like a select statement, except that a delete clause is used instead of a select clause. If you are deleting rows from a single table using a multitable delete format, the difference becomes even less noticeable. For example, here's a select statement that finds the account IDs of all accounts owned by John Hayward:

```
mysql> SELECT account2.account_id
    -> FROM account2 INNER JOIN customer2
    ->   ON account2.cust_id = customer2.cust_id
    ->   INNER JOIN individual2
    ->   ON individual2.cust_id = customer2.cust_id
    -> WHERE individual2.fname = 'John'
    ->   AND individual2.lname = 'Hayward';
+------------+
| account_id |
+------------+
|          8 |
|          9 |
|         10 |
+------------+
3 rows in set (0.01 sec)
```

If, after viewing the results, you decide to delete all three of John's accounts from the account2 table, you need only replace the select clause in the previous query with a delete clause naming the account2 table, as in:

```
mysql> DELETE account2
    -> FROM account2 INNER JOIN customer2
    ->   ON account2.cust_id = customer2.cust_id
    ->   INNER JOIN individual2
    ->   ON customer2.cust_id = individual2.cust_id
```

```
        -> WHERE individual2.fname = 'John'
        ->    AND individual2.lname = 'Hayward';
Query OK, 3 rows affected (0.01 sec)
```

Hopefully, this gives you a better idea of what the `delete` and `from` clauses are used for in a multitable `delete` statement. This statement is functionally identical to the following single-table `delete` statement, which uses a subquery to identify the customer ID of John Hayward:

```
DELETE FROM account2
WHERE cust_id =
 (SELECT cust_id
  FROM individual2
  WHERE fname = 'John' AND lname = 'Hayward';
```

When using a multitable `delete` statement to delete rows from a single table, you are simply choosing to use a querylike format involving table joins rather than a traditional `delete` statement using subqueries. The real power of multitable `delete` statements lies in the ability to delete from multiple tables in a single statement, as I demonstrated in the first statement in this section.

Along with the ability to delete rows from multiple tables, MySQL also gives you the ability to *modify* rows in multiple tables using a *multitable update*. Let's say that your bank is merging with another bank, and the databases from both banks have overlapping customer IDs. Your management decides to fix the problem by incrementing each customer ID in your database by 10,000 so that the second bank's data can be safely imported. The following statement shows how to modify the ID of customer ID 3 across the `individual2`, `customer2`, and `account2` tables using a single statement:

```
mysql> UPDATE individual2 INNER JOIN customer2
    ->    ON individual2.cust_id = customer2.cust_id
    ->    INNER JOIN account2
    ->    ON customer2.cust_id = account2.cust_id
    -> SET individual2.cust_id = individual2.cust_id + 10000,
    ->    customer2.cust_id = customer2.cust_id + 10000,
    ->    account2.cust_id = account2.cust_id + 10000
    -> WHERE individual2.cust_id = 3;
Query OK, 4 rows affected (0.01 sec)
Rows matched: 5  Changed: 4  Warnings: 0
```

This statement modifies four rows: one in each of the `individual2` and `customer2` tables, and two in the `account2` table. The multitable `update` syntax is very similar to that of the single-table `update`, except that the `update` clause contains multiple tables and their corresponding join conditions rather than just naming a single table. Just like the single-table `update`, the multitable version includes a `set` clause, the difference being that any tables referenced in the `update` clause may be modified via the `set` clause.

If you are using the InnoDB storage engine, you will most likely not be able to use multitable `delete` and `update` statements if the tables involved have foreign key constraints. This is because the engine does not guarantee that the changes will be applied in an order that won't violate the constraints. Instead, you should use multiple single-table statements in the proper order so that foreign key constraints are not violated.

Solutions to Exercises

Chapter 3

3-1

Retrieve the employee ID, first name, and last name for all bank employees. Sort by last name and then by first name.

```
mysql> SELECT emp_id, fname, lname
    -> FROM employee
    -> ORDER BY lname, fname;
+--------+----------+-----------+
| emp_id | fname    | lname     |
+--------+----------+-----------+
|      2 | Susan    | Barker    |
|     13 | John     | Blake     |
|      6 | Helen    | Fleming   |
|     17 | Beth     | Fowler    |
|      5 | John     | Gooding   |
|      9 | Jane     | Grossman  |
|      4 | Susan    | Hawthorne |
|     12 | Samantha | Jameson   |
|     16 | Theresa  | Markham   |
|     14 | Cindy    | Mason     |
|      8 | Sarah    | Parker    |
|     15 | Frank    | Portman   |
|     10 | Paula    | Roberts   |
|      1 | Michael  | Smith     |
|      7 | Chris    | Tucker    |
|     18 | Rick     | Tulman    |
|      3 | Robert   | Tyler     |
|     11 | Thomas   | Ziegler   |
+--------+----------+-----------+
18 rows in set (0.01 sec)
```

3-2

Retrieve the account ID, customer ID, and available balance for all accounts whose status equals 'ACTIVE' and whose available balance is greater than $2,500.

```
mysql> SELECT account_id, cust_id, avail_balance
    -> FROM account
    -> WHERE status = 'ACTIVE'
    ->   AND avail_balance > 2500;
+------------+---------+---------------+
| account_id | cust_id | avail_balance |
+------------+---------+---------------+
|          3 |       1 |       3000.00 |
|         10 |       4 |       5487.09 |
|         13 |       6 |      10000.00 |
|         14 |       7 |       5000.00 |
|         15 |       8 |       3487.19 |
|         18 |       9 |       9345.55 |
|         20 |      10 |      23575.12 |
|         22 |      11 |       9345.55 |
|         23 |      12 |      38552.05 |
|         24 |      13 |      50000.00 |
+------------+---------+---------------+
10 rows in set (0.00 sec)
```

3-3

Write a query against the account table that returns the IDs of the employees who opened the accounts (use the account.open_emp_id column). Include a single row for each distinct employee.

```
mysql> SELECT DISTINCT open_emp_id
    -> FROM account;
+-------------+
| open_emp_id |
+-------------+
|           1 |
|          10 |
|          13 |
|          16 |
+-------------+
4 rows in set (0.00 sec)
```

3-4

Fill in the blanks (denoted by <#>) for this multi-data-set query to achieve the results shown here:

```
mysql> SELECT p.product_cd, a.cust_id, a.avail_balance
    -> FROM product p INNER JOIN account <1>
    ->   ON p.product_cd = <2>
    -> WHERE p.<3> = 'ACCOUNT';
```

```
+------------+---------+---------------+
| product_cd | cust_id | avail_balance |
+------------+---------+---------------+
| CD         |       1 |       3000.00 |
| CD         |       6 |      10000.00 |
| CD         |       7 |       5000.00 |
| CD         |       9 |       1500.00 |
| CHK        |       1 |       1057.75 |
| CHK        |       2 |       2258.02 |
| CHK        |       3 |       1057.75 |
| CHK        |       4 |        534.12 |
| CHK        |       5 |       2237.97 |
| CHK        |       6 |        122.37 |
| CHK        |       8 |       3487.19 |
| CHK        |       9 |        125.67 |
| CHK        |      10 |      23575.12 |
| CHK        |      12 |      38552.05 |
| MM         |       3 |       2212.50 |
| MM         |       4 |       5487.09 |
| MM         |       9 |       9345.55 |
| SAV        |       1 |        500.00 |
| SAV        |       2 |        200.00 |
| SAV        |       4 |        767.77 |
| SAV        |       8 |        387.99 |
+------------+---------+---------------+
21 rows in set (0.02 sec)
```

The correct values for <1>, <2>, and <3> are:

1. a
2. a.product_cd
3. product_type_cd

Chapter 4

4-1

Which of the transaction IDs would be returned by the following filter conditions?

```
txn_date < '2005-02-26' AND (txn_type_cd = 'DBT' OR amount > 100)
```

Transaction IDs 1, 2, 3, 5, 6, and 7.

4-2

Which of the transaction IDs would be returned by the following filter conditions?

```
account_id IN (101,103) AND NOT (txn_type_cd = 'DBT' OR amount > 100)
```

Transaction IDs 4 and 9.

4-3

Construct a query that retrieves all accounts opened in 2002.

```
mysql> SELECT account_id, open_date
    -> FROM account
    -> WHERE open_date BETWEEN '2002-01-01' AND '2002-12-31';
+------------+------------+
| account_id | open_date  |
+------------+------------+
|          6 | 2002-11-23 |
|          7 | 2002-12-15 |
|         12 | 2002-08-24 |
|         20 | 2002-09-30 |
|         21 | 2002-10-01 |
+------------+------------+
5 rows in set (0.01 sec)
```

4-4

Construct a query that finds all nonbusiness customers whose last name contains an *a* in the second position and an *e* anywhere after the *a*.

```
mysql> SELECT cust_id, lname, fname
    -> FROM individual
    -> WHERE lname LIKE '_a%e%';
+---------+--------+---------+
| cust_id | lname  | fname   |
+---------+--------+---------+
|       1 | Hadley | James   |
|       9 | Farley | Richard |
+---------+--------+---------+
2 rows in set (0.02 sec)
```

Chapter 5

5-1

Fill in the blanks (denoted by <#>) for the following query to obtain the results that follow:

```
mysql> SELECT e.emp_id, e.fname, e.lname, b.name
    -> FROM employee e INNER JOIN <1> b
    ->   ON e.assigned_branch_id = b.<2>;
+--------+----------+-----------+----------------+
| emp_id | fname    | lname     | name           |
+--------+----------+-----------+----------------+
|      1 | Michael  | Smith     | Headquarters   |
|      2 | Susan    | Barker    | Headquarters   |
|      3 | Robert   | Tyler     | Headquarters   |
|      4 | Susan    | Hawthorne | Headquarters   |
|      5 | John     | Gooding   | Headquarters   |
```

```
|   6 | Helen    | Fleming  | Headquarters  |
|   7 | Chris    | Tucker   | Headquarters  |
|   8 | Sarah    | Parker   | Headquarters  |
|   9 | Jane     | Grossman | Headquarters  |
|  10 | Paula    | Roberts  | Woburn Branch |
|  11 | Thomas   | Ziegler  | Woburn Branch |
|  12 | Samantha | Jameson  | Woburn Branch |
|  13 | John     | Blake    | Quincy Branch |
|  14 | Cindy    | Mason    | Quincy Branch |
|  15 | Frank    | Portman  | Quincy Branch |
|  16 | Theresa  | Markham  | So. NH Branch |
|  17 | Beth     | Fowler   | So. NH Branch |
|  18 | Rick     | Tulman   | So. NH Branch |
+--------+----------+----------+---------------+
18 rows in set (0.03 sec)
```

The correct values for `<1>` and `<2>` are:

1. branch
2. branch_id

5-2

Write a query that returns the account ID for each nonbusiness customer (customer.cust_type_cd = 'I') along with the customer's federal ID (customer.fed_id) and the name of the product on which the account is based (product.name).

```
mysql> SELECT a.account_id, c.fed_id, p.name
    -> FROM account a INNER JOIN customer c
    ->   ON a.cust_id = c.cust_id
    ->   INNER JOIN product p
    ->   ON a.product_cd = p.product_cd
    -> WHERE c.cust_type_cd = 'I';
+------------+-------------+------------------------+
| account_id | fed_id      | name                   |
+------------+-------------+------------------------+
|          1 | 111-11-1111 | checking account       |
|          2 | 111-11-1111 | savings account        |
|          3 | 111-11-1111 | certificate of deposit |
|          4 | 222-22-2222 | checking account       |
|          5 | 222-22-2222 | savings account        |
|          6 | 333-33-3333 | checking account       |
|          7 | 333-33-3333 | money market account   |
|          8 | 444-44-4444 | checking account       |
|          9 | 444-44-4444 | savings account        |
|         10 | 444-44-4444 | money market account   |
|         11 | 555-55-5555 | checking account       |
|         12 | 666-66-6666 | checking account       |
|         13 | 666-66-6666 | certificate of deposit |
|         14 | 777-77-7777 | certificate of deposit |
|         15 | 888-88-8888 | checking account       |
|         16 | 888-88-8888 | savings account        |
|         17 | 999-99-9999 | checking account       |
```

```
|            18 | 999-99-9999 | money market account  |
|            19 | 999-99-9999 | certificate of deposit |
+------------+-------------+------------------------+
19 rows in set (0.00 sec)
```

5-3

Construct a query that finds all employees whose supervisor is assigned to a different department. Retrieve the employees' ID, first name, and last name.

```
mysql> SELECT e.emp_id, e.fname, e.lname
    -> FROM employee e INNER JOIN employee mgr
    ->   ON e.superior_emp_id = mgr.emp_id
    -> WHERE e.dept_id != mgr.dept_id;
+--------+-------+-----------+
| emp_id | fname | lname     |
+--------+-------+-----------+
|      4 | Susan | Hawthorne |
|      5 | John  | Gooding   |
+--------+-------+-----------+
2 rows in set (0.00 sec)
```

Chapter 6

6-1

If set A = {L M N O P} and set B = {P Q R S T}, what sets are generated by the following operations?

- A union B
- A union all B
- A intersect B
- A except B

1. A union B = {L M N O P Q R S T}
2. A union all B = {L M N O P P Q R S T}
3. A intersect B = {P}
4. A except B = {L M N O}

6-2

Write a compound query that finds the first and last names of all individual customers along with the first and last names of all employees.

```
mysql> SELECT fname, lname
    -> FROM individual
    -> UNION
```

```
    -> SELECT fname, lname
    -> FROM employee;
+----------+----------+
| fname    | lname    |
+----------+----------+
| James    | Hadley    |
| Susan    | Tingley   |
| Frank    | Tucker    |
| John     | Hayward   |
| Charles  | Frasier   |
| John     | Spencer   |
| Margaret | Young     |
| Louis    | Blake     |
| Richard  | Farley    |
| Michael  | Smith     |
| Susan    | Barker    |
| Robert   | Tyler     |
| Susan    | Hawthorne |
| John     | Gooding   |
| Helen    | Fleming   |
| Chris    | Tucker    |
| Sarah    | Parker    |
| Jane     | Grossman  |
| Paula    | Roberts   |
| Thomas   | Ziegler   |
| Samantha | Jameson   |
| John     | Blake     |
| Cindy    | Mason     |
| Frank    | Portman   |
| Theresa  | Markham   |
| Beth     | Fowler    |
| Rick     | Tulman    |
+----------+----------+
27 rows in set (0.01 sec)
```

6-3

Sort the results from Exercise 6-2 by the lname column.

```
mysql> SELECT fname, lname
    -> FROM individual
    -> UNION ALL
    -> SELECT fname, lname
    -> FROM employee
    -> ORDER BY lname;
+----------+----------+
| fname    | lname    |
+----------+----------+
| Susan    | Barker   |
| Louis    | Blake    |
| John     | Blake    |
| Richard  | Farley   |
| Helen    | Fleming  |
| Beth     | Fowler   |
| Charles  | Frasier  |
```

```
| John     | Gooding   |
| Jane     | Grossman  |
| James    | Hadley    |
| Susan    | Hawthorne |
| John     | Hayward   |
| Samantha | Jameson   |
| Theresa  | Markham   |
| Cindy    | Mason     |
| Sarah    | Parker    |
| Frank    | Portman   |
| Paula    | Roberts   |
| Michael  | Smith     |
| John     | Spencer   |
| Susan    | Tingley   |
| Chris    | Tucker    |
| Frank    | Tucker    |
| Rick     | Tulman    |
| Robert   | Tyler     |
| Margaret | Young     |
| Thomas   | Ziegler   |
+----------+-----------+
27 rows in set (0.01 sec)
```

Chapter 7

7-1

Write a query that returns the 17th through 25th characters of the string `'Please find the substring in this string'`.

```
mysql> SELECT SUBSTRING('Please find the substring in this string',17,9);
+-----------------------------------------------------------+
| SUBSTRING('Please find the substring in this string',17,9) |
+-----------------------------------------------------------+
| substring                                                 |
+-----------------------------------------------------------+
1 row in set (0.00 sec)
```

7-2

Write a query that returns the absolute value and sign (-1, 0, or 1) of the number -25. 76823. Also return the number rounded to the nearest hundredth.

```
mysql> SELECT ABS(-25.76823), SIGN(-25.76823), ROUND(-25.76823, 2);
+----------------+-----------------+---------------------+
| ABS(-25.76823) | SIGN(-25.76823) | ROUND(-25.76823, 2) |
+----------------+-----------------+---------------------+
|       25.76823 |              -1 |              -25.77 |
+----------------+-----------------+---------------------+
1 row in set (0.00 sec)
```

7-3

Write a query to return just the month portion of the current date.

```
mysql> SELECT EXTRACT(MONTH FROM CURRENT_DATE());
+-----------------------------------+
| EXTRACT(MONTH FROM CURRENT_DATE) |
+-----------------------------------+
|                                 5 |
+-----------------------------------+
1 row in set (0.02 sec)
```

(Your result will most likely be different, unless it happens to be May when you try this exercise.)

Chapter 8

8-1

Construct a query that counts the number of rows in the account table.

```
mysql> SELECT COUNT(*)
    -> FROM account;
+----------+
| count(*) |
+----------+
|       24 |
+----------+
1 row in set (0.32 sec)
```

8-2

Modify your query from Exercise 8-1 to count the number of accounts held by each customer. Show the customer ID and the number of accounts for each customer.

```
mysql> SELECT cust_id, COUNT(*)
    -> FROM account
    -> GROUP BY cust_id;
+---------+----------+
| cust_id | count(*) |
+---------+----------+
|       1 |        3 |
|       2 |        2 |
|       3 |        2 |
|       4 |        3 |
|       5 |        1 |
|       6 |        2 |
|       7 |        1 |
|       8 |        2 |
|       9 |        3 |
|      10 |        2 |
|      11 |        1 |
```

```
|      12 |        1 |
|      13 |        1 |
+---------+----------+
13 rows in set (0.00 sec)
```

8-3

Modify your query from Exercise 8-2 to include only those customers having at least two accounts.

```
mysql> SELECT cust_id, COUNT(*)
    -> FROM account
    -> GROUP BY cust_id
    -> HAVING COUNT(*) >= 2;
+---------+----------+
| cust_id | COUNT(*) |
+---------+----------+
|       1 |        3 |
|       2 |        2 |
|       3 |        2 |
|       4 |        3 |
|       6 |        2 |
|       8 |        2 |
|       9 |        3 |
|      10 |        2 |
+---------+----------+
8 rows in set (0.04 sec)
```

8-4 (Extra Credit)

Find the total available balance by product and branch where there is more than one account per product and branch. Order the results by total balance (highest to lowest).

```
mysql> SELECT product_cd, open_branch_id, SUM(avail_balance)
    -> FROM account
    -> GROUP BY product_cd, open_branch_id
    -> HAVING COUNT(*) > 1
    -> ORDER BY 3 DESC;
+------------+----------------+--------------------+
| product_cd | open_branch_id | SUM(avail_balance) |
+------------+----------------+--------------------+
| CHK        |              4 |           67852.33 |
| MM         |              1 |           14832.64 |
| CD         |              1 |           11500.00 |
| CD         |              2 |            8000.00 |
| CHK        |              2 |            3315.77 |
| CHK        |              1 |             782.16 |
| SAV        |              2 |             700.00 |
+------------+----------------+--------------------+
7 rows in set (0.01 sec)
```

Note that MySQL would not accept ORDER BY SUM(avail_balance) DESC,, so I was forced to indicate the sort column by position.

Chapter 9

9-1

Construct a query against the account table that uses a filter condition with a noncorrelated subquery against the product table to find all loan accounts (product.product_type_cd = 'LOAN'). Retrieve the account ID, product code, customer ID, and available balance.

```
mysql> SELECT account_id, product_cd, cust_id, avail_balance
    -> FROM account
    -> WHERE product_cd IN (SELECT product_cd
    ->    FROM product
    ->    WHERE product_type_cd = 'LOAN');
+------------+------------+---------+---------------+
| account_id | product_cd | cust_id | avail_balance |
+------------+------------+---------+---------------+
|         21 | BUS        |      10 |          0.00 |
|         22 | BUS        |      11 |       9345.55 |
|         24 | SBL        |      13 |      50000.00 |
+------------+------------+---------+---------------+
3 rows in set (0.07 sec)
```

9-2

Rework the query from Exercise 9-1 using a *correlated* subquery against the product table to achieve the same results.

```
mysql> SELECT a.account_id, a.product_cd, a.cust_id, a.avail_balance
    -> FROM account a
    -> WHERE EXISTS (SELECT 1
    ->    FROM product p
    ->    WHERE p.product_cd = a.product_cd
    ->       AND p.product_type_cd = 'LOAN');
+------------+------------+---------+---------------+
| account_id | product_cd | cust_id | avail_balance |
+------------+------------+---------+---------------+
|         21 | BUS        |      10 |          0.00 |
|         22 | BUS        |      11 |       9345.55 |
|         24 | SBL        |      13 |      50000.00 |
+------------+------------+---------+---------------+
3 rows in set (0.01 sec)
```

9-3

Join the following query to the employee table to show the experience level of each employee:

```
SELECT 'trainee' name, '2004-01-01' start_dt, '2005-12-31' end_dt
UNION ALL
SELECT 'worker' name, '2002-01-01' start_dt, '2003-12-31' end_dt
```

```
                           UNION ALL
                           SELECT 'mentor' name, '2000-01-01' start_dt, '2001-12-31' end_dt
```

Give the subquery the alias `levels`, and include the employee's ID, first name, last name, and experience level (`levels.name`). (Hint: build a join condition using an inequality condition to determine into which level the `employee.start_date` column falls.)

```
mysql> SELECT e.emp_id, e.fname, e.lname, levels.name
    -> FROM employee e INNER JOIN
    -> (SELECT 'trainee' name, '2004-01-01' start_dt, '2005-12-31' end_dt
    ->   UNION ALL
    -> SELECT 'worker' name, '2002-01-01' start_dt, '2003-12-31' end_dt
    ->   UNION ALL
    -> SELECT 'mentor' name, '2000-01-01' start_dt, '2001-12-31' end_dt) levels
    -> ON e.start_date BETWEEN levels.start_dt AND levels.end_dt;
```

```
+--------+----------+-----------+---------+
| emp_id | fname    | lname     | name    |
+--------+----------+-----------+---------+
|      6 | Helen    | Fleming   | trainee |
|      7 | Chris    | Tucker    | trainee |
|      2 | Susan    | Barker    | worker  |
|      4 | Susan    | Hawthorne | worker  |
|      5 | John     | Gooding   | worker  |
|      8 | Sarah    | Parker    | worker  |
|      9 | Jane     | Grossman  | worker  |
|     10 | Paula    | Roberts   | worker  |
|     12 | Samantha | Jameson   | worker  |
|     14 | Cindy    | Mason     | worker  |
|     15 | Frank    | Portman   | worker  |
|     17 | Beth     | Fowler    | worker  |
|     18 | Rick     | Tulman    | worker  |
|      1 | Michael  | Smith     | mentor  |
|      3 | Robert   | Tyler     | mentor  |
|     11 | Thomas   | Ziegler   | mentor  |
|     13 | John     | Blake     | mentor  |
|     16 | Theresa  | Markham   | mentor  |
+--------+----------+-----------+---------+
18 rows in set (0.00 sec)
```

9-4

Construct a query against the `employee` table that retrieves the employee ID, first name, and last name, along with the names of the department and branch to which the employee is assigned. Do not join any tables.

```
mysql> SELECT e.emp_id, e.fname, e.lname,
    -> (SELECT d.name FROM department d
    ->  WHERE d.dept_id = e.dept_id) dept_name,
    -> (SELECT b.name FROM branch b
    ->  WHERE b.branch_id = e.assigned_branch_id) branch_name
    -> FROM employee e;
+--------+----------+-----------+----------------+---------------+
```

```
| emp_id | fname    | lname     | dept_name      | branch_name     |
+--------+----------+-----------+----------------+-----------------+
|      1 | Michael  | Smith     | Administration | Headquarters    |
|      2 | Susan    | Barker    | Administration | Headquarters    |
|      3 | Robert   | Tyler     | Administration | Headquarters    |
|      4 | Susan    | Hawthorne | Operations     | Headquarters    |
|      5 | John     | Gooding   | Loans          | Headquarters    |
|      6 | Helen    | Fleming   | Operations     | Headquarters    |
|      7 | Chris    | Tucker    | Operations     | Headquarters    |
|      8 | Sarah    | Parker    | Operations     | Headquarters    |
|      9 | Jane     | Grossman  | Operations     | Headquarters    |
|     10 | Paula    | Roberts   | Operations     | Woburn Branch   |
|     11 | Thomas   | Ziegler   | Operations     | Woburn Branch   |
|     12 | Samantha | Jameson   | Operations     | Woburn Branch   |
|     13 | John     | Blake     | Operations     | Quincy Branch   |
|     14 | Cindy    | Mason     | Operations     | Quincy Branch   |
|     15 | Frank    | Portman   | Operations     | Quincy Branch   |
|     16 | Theresa  | Markham   | Operations     | So. NH Branch   |
|     17 | Beth     | Fowler    | Operations     | So. NH Branch   |
|     18 | Rick     | Tulman    | Operations     | So. NH Branch   |
+--------+----------+-----------+----------------+-----------------+
18 rows in set (0.12 sec)
```

Chapter 10

10-1

Write a query that returns all product names along with the accounts based on that product (use the product_cd column in the account table to link to the product table). Include all products, even if no accounts have been opened for that product.

```
mysql> SELECT p.product_cd, a.account_id, a.cust_id, a.avail_balance
    -> FROM product p LEFT OUTER JOIN account a
    ->   ON p.product_cd = a.product_cd;
+------------+------------+---------+---------------+
| product_cd | account_id | cust_id | avail_balance |
+------------+------------+---------+---------------+
| AUT        |       NULL |    NULL |          NULL |
| BUS        |         21 |      10 |          0.00 |
| BUS        |         22 |      11 |       9345.55 |
| CD         |          3 |       1 |       3000.00 |
| CD         |         13 |       6 |      10000.00 |
| CD         |         14 |       7 |       5000.00 |
| CD         |         19 |       9 |       1500.00 |
| CHK        |          1 |       1 |       1057.75 |
| CHK        |          4 |       2 |       2258.02 |
| CHK        |          6 |       3 |       1057.75 |
| CHK        |          8 |       4 |        534.12 |
| CHK        |         11 |       5 |       2237.97 |
| CHK        |         12 |       6 |        122.37 |
| CHK        |         15 |       8 |       3487.19 |
| CHK        |         17 |       9 |        125.67 |
| CHK        |         20 |      10 |      23575.12 |
```

```
| CHK     |          23 |      12 |       38552.05 |
| MM      |           7 |       3 |        2212.50 |
| MM      |          10 |       4 |        5487.09 |
| MM      |          18 |       9 |        9345.55 |
| MRT     |        NULL |    NULL |           NULL |
| SAV     |           2 |       1 |         500.00 |
| SAV     |           5 |       2 |         200.00 |
| SAV     |           9 |       4 |         767.77 |
| SAV     |          16 |       8 |         387.99 |
| SBL     |          24 |      13 |       50000.00 |
+---------+-------------+---------+----------------+
26 rows in set (0.01 sec)
```

10-2

Reformulate your query from Exercise 10-1 to use the other outer join type (e.g., if you used a left outer join in Exercise 10-1, use a right outer join this time) such that the results are identical to Exercise 10-1.

```
mysql> SELECT p.product_cd, a.account_id, a.cust_id, a.avail_balance
    -> FROM account a RIGHT OUTER JOIN product p
    ->   ON p.product_cd = a.product_cd;
+------------+-------------+---------+----------------+
| product_cd | account_id  | cust_id | avail_balance  |
+------------+-------------+---------+----------------+
| AUT        |        NULL |    NULL |           NULL |
| BUS        |          21 |      10 |           0.00 |
| BUS        |          22 |      11 |        9345.55 |
| CD         |           3 |       1 |        3000.00 |
| CD         |          13 |       6 |       10000.00 |
| CD         |          14 |       7 |        5000.00 |
| CD         |          19 |       9 |        1500.00 |
| CHK        |           1 |       1 |        1057.75 |
| CHK        |           4 |       2 |        2258.02 |
| CHK        |           6 |       3 |        1057.75 |
| CHK        |           8 |       4 |         534.12 |
| CHK        |          11 |       5 |        2237.97 |
| CHK        |          12 |       6 |         122.37 |
| CHK        |          15 |       8 |        3487.19 |
| CHK        |          17 |       9 |         125.67 |
| CHK        |          20 |      10 |       23575.12 |
| CHK        |          23 |      12 |       38552.05 |
| MM         |           7 |       3 |        2212.50 |
| MM         |          10 |       4 |        5487.09 |
| MM         |          18 |       9 |        9345.55 |
| MRT        |        NULL |    NULL |           NULL |
| SAV        |           2 |       1 |         500.00 |
| SAV        |           5 |       2 |         200.00 |
| SAV        |           9 |       4 |         767.77 |
| SAV        |          16 |       8 |         387.99 |
| SBL        |          24 |      13 |       50000.00 |
+------------+-------------+---------+----------------+
26 rows in set (0.02 sec)
```

10-3

Outer-join the `account` table to both the `individual` and `business` tables (via the `account.cust_id` column) such that the result set contains one row per account. Columns to include are account.account_id, account.product_cd, individual.fname, individual.lname, and business.name.

```
mysql> SELECT a.account_id, a.product_cd,
    ->   i.fname, i.lname, b.name
    -> FROM account a LEFT OUTER JOIN business b
    ->   ON a.cust_id = b.cust_id
    ->   LEFT OUTER JOIN individual i
    ->   ON a.cust_id = i.cust_id;
```

account_id	product_cd	fname	lname	name
1	CHK	James	Hadley	NULL
2	SAV	James	Hadley	NULL
3	CD	James	Hadley	NULL
4	CHK	Susan	Tingley	NULL
5	SAV	Susan	Tingley	NULL
6	CHK	Frank	Tucker	NULL
7	MM	Frank	Tucker	NULL
8	CHK	John	Hayward	NULL
9	SAV	John	Hayward	NULL
10	MM	John	Hayward	NULL
11	CHK	Charles	Frasier	NULL
12	CHK	John	Spencer	NULL
13	CD	John	Spencer	NULL
14	CD	Margaret	Young	NULL
15	CHK	Louis	Blake	NULL
16	SAV	Louis	Blake	NULL
17	CHK	Richard	Farley	NULL
18	MM	Richard	Farley	NULL
19	CD	Richard	Farley	NULL
20	CHK	NULL	NULL	Chilton Engineering
21	BUS	NULL	NULL	Chilton Engineering
22	BUS	NULL	NULL	Northeast Cooling Inc.
23	CHK	NULL	NULL	Superior Auto Body
24	SBL	NULL	NULL	AAA Insurance Inc.

```
24 rows in set (0.05 sec)
```

10-4 (Extra Credit)

Devise a query that will generate the set {1, 2, 3,..., 99, 100}. (Hint: use a cross join with at least two `from` clause subqueries.)

```
SELECT ones.x + tens.x + 1
FROM
 (SELECT 0 x UNION ALL
  SELECT 1 x UNION ALL
  SELECT 2 x UNION ALL
  SELECT 3 x UNION ALL
```

```
      SELECT 4 x UNION ALL
      SELECT 5 x UNION ALL
      SELECT 6 x UNION ALL
      SELECT 7 x UNION ALL
      SELECT 8 x UNION ALL
      SELECT 9 x) ones
CROSS JOIN
(SELECT 0 x UNION ALL
      SELECT 10 x UNION ALL
      SELECT 20 x UNION ALL
      SELECT 30 x UNION ALL
      SELECT 40 x UNION ALL
      SELECT 50 x UNION ALL
      SELECT 60 x UNION ALL
      SELECT 70 x UNION ALL
      SELECT 80 x UNION ALL
      SELECT 90 x) tens;
```

Chapter 11

11-1

Rewrite the following query, which uses a simple case expression, so that the same results are achieved using a searched case expression. Try to use as few when clauses as possible.

```
SELECT emp_id,
  CASE title
    WHEN 'President' THEN 'Management'
    WHEN 'Vice President' THEN 'Management'
    WHEN 'Treasurer' THEN 'Management'
    WHEN 'Loan Manager' THEN 'Management'
    WHEN 'Operations Manager' THEN 'Operations'
    WHEN 'Head Teller' THEN 'Operations'
    WHEN 'Teller' THEN 'Operations'
    ELSE 'Unknown'
  END
FROM employee;

 SELECT emp_id,
  CASE
    WHEN title LIKE '%President' OR title = 'Loan Manager'
      OR title = 'Treasurer'
      THEN 'Management'
    WHEN title LIKE '%Teller' OR title = 'Operations Manager'
      THEN 'Operations'
    ELSE 'Unknown'
  END
FROM employee;
```

11-2

Rewrite the following query so that the result set contains a single row with four columns (one for each branch). Name the four columns branch_1 through branch_4.

```
mysql> SELECT open_branch_id, COUNT(*)
    -> FROM account
    -> GROUP BY open_branch_id;
+----------------+----------+
| open_branch_id | COUNT(*) |
+----------------+----------+
|              1 |        8 |
|              2 |        7 |
|              3 |        3 |
|              4 |        6 |
+----------------+----------+
4 rows in set (0.00 sec)

mysql> SELECT
    ->    SUM(CASE WHEN open_branch_id = 1 THEN 1 ELSE 0 END) branch_1,
    ->    SUM(CASE WHEN open_branch_id = 2 THEN 1 ELSE 0 END) branch_2,
    ->    SUM(CASE WHEN open_branch_id = 3 THEN 1 ELSE 0 END) branch_3,
    ->    SUM(CASE WHEN open_branch_id = 4 THEN 1 ELSE 0 END) branch_4
    -> FROM account;
+----------+----------+----------+----------+
| branch_1 | branch_2 | branch_3 | branch_4 |
+----------+----------+----------+----------+
|        8 |        7 |        3 |        6 |
+----------+----------+----------+----------+
1 row in set (0.02 sec)
```

Chapter 12

12-1

Generate a transaction to transfer $50 from Frank Tucker's money market account to his checking account. You will need to insert two rows into the transaction table and update two rows in the account table.

```
START TRANSACTION;

SELECT i.cust_id,
  (SELECT a.account_id FROM account a
   WHERE a.cust_id = i.cust_id
     AND a.product_cd = 'MM') mm_id,
  (SELECT a.account_id FROM account a
   WHERE a.cust_id = i.cust_id
     AND a.product_cd = 'chk') chk_id
INTO @cst_id, @mm_id, @chk_id
FROM individual i
WHERE i.fname = 'Frank' AND i.lname = 'Tucker';
```

```
INSERT INTO transaction (txn_id, txn_date, account_id,
  txn_type_cd, amount)
VALUES (NULL, now(), @mm_id, 'CDT', 50);

INSERT INTO transaction (txn_id, txn_date, account_id,
  txn_type_cd, amount)
VALUES (NULL, now(), @chk_id, 'DBT', 50);

UPDATE account
SET last_activity_date = now(),
  avail_balance = avail_balance - 50
WHERE account_id = @mm_id;

UPDATE account
SET last_activity_date = now(),
  avail_balance = avail_balance + 50
WHERE account_id = @chk_id;

COMMIT;
```

Chapter 13

13-1

Modify the account table so that customers may not have more than one account for each product.

```
ALTER TABLE account
ADD CONSTRAINT account_unq1 UNIQUE (cust_id, product_cd);
```

13-2

Generate a multicolumn index on the transaction table that could be used by both of the following queries:

```
SELECT txn_date, account_id, txn_type_cd, amount
FROM transaction
WHERE txn_date > cast('2008-12-31 23:59:59' as datetime);

SELECT txn_date, account_id, txn_type_cd, amount
FROM transaction
WHERE txn_date > cast('2008-12-31 23:59:59' as datetime)
  AND amount < 1000;

CREATE INDEX txn_idx01
ON transaction (txn_date, amount);
```

Chapter 14

14-1

Create a view that queries the `employee` table and generates the following output when queried with no `where` clause:

```
+------------------+------------------+
| supervisor_name  | employee_name    |
+------------------+------------------+
| NULL             | Michael Smith    |
| Michael Smith    | Susan Barker     |
| Michael Smith    | Robert Tyler     |
| Robert Tyler     | Susan Hawthorne  |
| Susan Hawthorne  | John Gooding     |
| Susan Hawthorne  | Helen Fleming    |
| Helen Fleming    | Chris Tucker     |
| Helen Fleming    | Sarah Parker     |
| Helen Fleming    | Jane Grossman    |
| Susan Hawthorne  | Paula Roberts    |
| Paula Roberts    | Thomas Ziegler   |
| Paula Roberts    | Samantha Jameson |
| Susan Hawthorne  | John Blake       |
| John Blake       | Cindy Mason      |
| John Blake       | Frank Portman    |
| Susan Hawthorne  | Theresa Markham  |
| Theresa Markham  | Beth Fowler      |
| Theresa Markham  | Rick Tulman      |
+------------------+------------------+
18 rows in set (1.47 sec)

mysql> CREATE VIEW supervisor_vw
    -> (supervisor_name,
    ->  employee_name
    -> )
    -> AS
    -> SELECT concat(spr.fname, ' ', spr.lname),
    ->   concat(emp.fname, ' ', emp.lname)
    -> FROM employee emp LEFT OUTER JOIN employee spr
    ->   ON emp.superior_emp_id = spr.emp_id;
Query OK, 0 rows affected (0.12 sec)

mysql> SELECT * FROM supervisor_vw;
+------------------+------------------+
| supervisor_name  | employee_name    |
+------------------+------------------+
| NULL             | Michael Smith    |
| Michael Smith    | Susan Barker     |
| Michael Smith    | Robert Tyler     |
| Robert Tyler     | Susan Hawthorne  |
| Susan Hawthorne  | John Gooding     |
| Susan Hawthorne  | Helen Fleming    |
| Helen Fleming    | Chris Tucker     |
```

```
| Helen Fleming    | Sarah Parker     |
| Helen Fleming    | Jane Grossman    |
| Susan Hawthorne  | Paula Roberts    |
| Paula Roberts    | Thomas Ziegler   |
| Paula Roberts    | Samantha Jameson |
| Susan Hawthorne  | John Blake       |
| John Blake       | Cindy Mason      |
| John Blake       | Frank Portman    |
| Susan Hawthorne  | Theresa Markham  |
| Theresa Markham  | Beth Fowler      |
| Theresa Markham  | Rick Tulman      |
+------------------+------------------+
18 rows in set (0.17 sec)
```

14-2

The bank president would like to have a report showing the name and city of each branch, along with the total balances of all accounts opened at the branch. Create a view to generate the data.

```
mysql> CREATE VIEW branch_summary_vw
    ->  (branch_name,
    ->   branch_city,
    ->   total_balance
    ->  )
    -> AS
    -> SELECT b.name, b.city, sum(a.avail_balance)
    -> FROM branch b INNER JOIN account a
    ->   ON b.branch_id = a.open_branch_id
    -> GROUP BY b.name, b.city;
Query OK, 0 rows affected (0.00 sec)

mysql> SELECT * FROM branch_summary_vw;
+----------------+-------------+---------------+
| branch_name    | branch_city | total_balance |
+----------------+-------------+---------------+
| Headquarters   | Waltham     |      27882.57 |
| Quincy Branch  | Quincy      |      53270.25 |
| So. NH Branch  | Salem       |      68240.32 |
| Woburn Branch  | Woburn      |      21361.32 |
+----------------+-------------+---------------+
4 rows in set (0.01 sec)
```

Chapter 15

15-1

Write a query that lists all the indexes in the bank schema. Include the table names.

```
mysql> SELECT DISTINCT table_name, index_name
    -> FROM information_schema.statistics
    -> WHERE table_schema = 'bank';
```

```
+--------------+---------------------+
| table_name   | index_name          |
+--------------+---------------------+
| account      | PRIMARY             |
| account      | account_unq1        |
| account      | fk_product_cd       |
| account      | fk_a_branch_id      |
| account      | fk_a_emp_id         |
| account      | acc_bal_idx         |
| branch       | PRIMARY             |
| business     | PRIMARY             |
| customer     | PRIMARY             |
| department   | PRIMARY             |
| department   | dept_name_idx       |
| employee     | PRIMARY             |
| employee     | fk_dept_id          |
| employee     | fk_e_branch_id      |
| employee     | fk_e_emp_id         |
| individual   | PRIMARY             |
| officer      | PRIMARY             |
| officer      | fk_o_cust_id        |
| product      | PRIMARY             |
| product      | fk_product_type_cd  |
| product_type | PRIMARY             |
| transaction  | PRIMARY             |
| transaction  | fk_t_account_id     |
| transaction  | fk_teller_emp_id    |
| transaction  | fk_exec_branch_id   |
| transaction  | txn_idx01           |
+--------------+---------------------+
26 rows in set (0.00 sec)
```

15-2

Write a query that generates output that can be used to create all of the indexes on the
bank.employee table. Output should be of the form:

```
"ALTER TABLE <table_name> ADD INDEX <index_name> (<column_list>)"
```

```
mysql> SELECT concat(
    ->   CASE
    ->     WHEN st.seq_in_index = 1 THEN
    ->       concat('ALTER TABLE ', st.table_name, ' ADD',
    ->         CASE
    ->           WHEN st.non_unique = 0 THEN ' UNIQUE '
    ->           ELSE ' '
    ->         END,
    ->         'INDEX ',
    ->         st.index_name, ' (', st.column_name)
    ->     ELSE concat(' ', st.column_name)
    ->   END,
    ->   CASE
    ->     WHEN st.seq_in_index =
    ->       (SELECT max(st2.seq_in_index)
    ->         FROM information_schema.statistics st2
```

```
    ->        WHERE st2.table_schema = st.table_schema
    ->          AND st2.table_name = st.table_name
    ->          AND st2.index_name = st.index_name)
    ->        THEN ');'
    ->      ELSE ''
    ->    END
    ->  ) index_creation_statement
    -> FROM information_schema.statistics st
    -> WHERE st.table_schema = 'bank'
    ->   AND st.table_name = 'employee'
    -> ORDER BY st.index_name, st.seq_in_index;
+---------------------------------------------------------------------+
| index_creation_statement                                           |
+---------------------------------------------------------------------+
| ALTER TABLE employee ADD INDEX fk_dept_id (dept_id);               |
| ALTER TABLE employee ADD INDEX fk_e_branch_id (assigned_branch_id); |
| ALTER TABLE employee ADD INDEX fk_e_emp_id (superior_emp_id);      |
| ALTER TABLE employee ADD UNIQUE INDEX PRIMARY (emp_id);            |
+---------------------------------------------------------------------+
4 rows in set (0.20 sec)
```

Index

Symbols

! (exclamation mark), != (not equal to)
    operator, 67
% (percent sign), wildcard character in partial
    string matches, 74
' ' (quotes, single)
    in strings, 116
    surrounding strings, 114
' (apostrophe) in strings, 116
( ) (parentheses)
    enclosing subqueries, 157
    ordering query combinations in compound
        queries, 111
    using with filter conditions, 64
< (less than) operator
    scalar subqueries and, 159
    using with all operator, 163
<= (less than or equal to) operator, 159
<> (not equal to) operator
    in inequality conditions, 67
    scalar subqueries and, 159
    using with all operator, 163
= (equals sign)
    = null, filtering for null values, 77
    equal to operator
        scalar subqueries and, 159
        using with all operator, 163
        using with any operator, 165
    in equality conditions, 67
> (greater than) operator
    scalar subqueries and, 159
    using with all operator, 163
>= (greater than or equal to) operator, 159

\ (backslash), escaping special characters in
    strings, 116
_ (underscore), wildcard character in partial
    string matches, 74

A

abs( ) function, 130
aggregate functions, 144, 145–150
    count( ) function, 147
    exercises with, 156
    handling null values, 149
    in having clause, 155
    implicit versus explicit groups, 146
    listing of common functions, 145
    using expressions as arguments, 149
    where clause and, 155
aggregation
    selective aggregation using case expressions,
        209
    using views for data aggregation, 249
all operator, 163
    <> all comparisons, null values and, 163
alter table statements
    adding or removing constraints, 239
    adding or removing indexes, 229
    changing storage engine, 224
    modifying definition of existing table, 31
and operator
    condition evaluation with, 63
    three-condition evaluation using and, or,
        64
    three-condition evaluation using and, or,
        and not, 65
    using in select statement where clause, 54
ANSI mode, 115

We'd like to hear your suggestions for improving our indexes. Send email to *index@oreilly.com*.

ANSI SQL standard, join syntax, 86
any operator, 165
arithmetic functions, 126
arithmetic operators in filter conditions, 66
as keyword
    using with column aliases, 47
    using with table aliases, 52
asc and desc keywords, 57
ASCII character set, 117
ascii( ) function, 118
atomicity, 219
auto-commit mode, 221
auto-increment feature in MySQL, 31
avg( ) function, 145

B

B-tree (balanced-tree) indexes, 232
begin transaction command, 221
between operator, 69
bitmap indexes, 233
branch nodes (B-tree indexes), 232

C

C language, SQL integration toolkits, 10
C#, SQL integration toolkit, 10
C++, SQL integration toolkits, 10
Cartesian products, 83, 192–198
cascading deletes, 242
cascading updates, 241
case expressions, 204
    examples, 207–215
        checking for existence, 211
        conditional updates, 214
        division-by-zero errors, 212
        handling null values, 214
        result set transformations, 208
        selective aggregation, 209
    searched, 205
    simple, 206
cast( ) function, 141
    converting strings to temporal data types,
        135
ceil( ) function, 128
char type, 19, 113
char( ) function, 117
    generating strings, 118
character data types, 113
character data, MySQL database, 18

character sets
    ASCII, 117
    latin1, 118
    sorting order or collation, 71
check constraints, 238
clauses
    referencing multiple tables joined in a query,
        52
    select statement, 43
        select clause, 43
    select, from, and where, 11
    subqueries in, 159
CLOB (Character Large Object) type, 114
code examples from this book, xii
collation, 71
column aliases, 46
columns, 5
    defined, 6
    incorrect values for, 37
    multicolumn grouping, 151
    query returning number of, 265
    single-column grouping, 151
    viewing for a table with describe command,
        39
columns view, 260
comments, 11
commit command, 219
    ending transactions, 222
    issuing for savepoints, 225
commits, auto-commit mode, 221
comparison operators
    in filter conditions, 66
    using scalar subqueries with, 159
    using with all operator, 163
    using with any operator, 165
complexity, hiding with use of views, 250
compound key, 5
compound queries, 103
concat( ) function, 118
    appending characters to strings, 123
    building strings from individual pieces of
        data, 123
conditional logic, 203–216
    case expressions, 204
        examples, 207–215
        searched, 205
        simple, 206
    defined, 203
    exercises in, 215

using correlated subqueries, 171
deletes, cascading, 242
deployment verification for schema objects, 265
desc and asc keywords, 57
describe (desc) command, 29
    columns in table, describing, 39
    examining views, 247
distinct keyword, 48, 162
division-by-zero errors, 212
driving table, 90
duplicates
    excluding using union operator, 105
    removal by except and except all operators, 108
    removal by intersect operator, 106
    removing from query returns, 47
    union all operator and, 104
durability, 220
dynamic SQL execution, 266

E

Eastern Standard Time, 131
entities, 5
    defined, 6
enum data type, MySQL, 28
equality conditions, 66
    case expressions and, 207
    correlated subqueries in, 167
    data modification using, 67
    error from subquery returning more than one row, 160
    subquery in, 158
equi-joins, 94
escaping single quotes in strings, 116
except all operator, 107
except operation (sets), 100
except operator, 107
execute statements, 267
existence, checking for, 211
exists operator, 169
explicit groups, 146
expressions, 204
    (see also case expressions)
    in filter conditions, 66
    generating with subqueries, 177
    grouping via, 152
    included in select clause (example), 45

sorting data in select statement order by clause, 58
using as arguments for aggregate functions, 149
extract( ) function, 139
    returning only year portion of a date, 152

F

Falcon storage engine, 224
filter conditions
    ANSI join syntax and, 86
    group, 145, 155
    join conditions versus, 96
    in select statement where clause, 52
    subqueries in, 177
    in where clauses, 63
filtering, 63–79
    building conditions, 66
    condition types, 66
        equality conditions, 66
        inequality conditions, 67
        matching conditions, 73
        membership conditions, 71
        modifying data with equality conditions, 67
        range conditions, 68
    evaluation of conditions, 63
        using not operator, 65
        using parentheses, 64
    exercises, 79
    null values, 76
floating-point numbers, controlling precision of, 128
floating-point types, MySQL, 22
floor( ) function, 128
foreign key constraints, 238
    cascading, 240–242
foreign keys, 6
    defined, 7
    nonexistent key causing statement errors, 36
    self-referencing, 93
    using in joins, 82
from clauses, 11
    join order and, 90
    joining three or more tables, 88
    joining two tables using inner join, 84
    missing, 18
    on subclause, 83

important considerations, 33
inserting data through views, 254
noncorrelated scalar subqueries generating
    values for, 180
insert( ) function, 124
integer types, MySQL, 22
integration toolkits for SQL, 9
intermediate result sets, 90
intersect all operator, 106
intersect operator, 106
    precedence of, 111
intersection operation (sets), 100
intervals
    adding to dates, 137
    common interval types, 138
    determing number between two dates, 140
    using with extract( ) function, 140
is not null operator, 77
is null operator, 76

J

Java
    SQL integration toolkits, 10
    SQL statements and, 266
join conditions
    ANSI join syntax and, 86
    filter conditions versus, 96
join keyword, 82
joins, 81, 183–201
    ANSI syntax for, 86
    cross joins, 83, 192–198
    defined, 82
    equi- versus non-equi-joins, 94–96
    exercises, 97
    exercises in, 200
    inner joins, 83
    join versus filter conditions, 96
    joining three or more tables, 88–93
        order of joins, 89
        specifying join order, 90
        using same table twice, 92
        using subqueries as tables, 90
    joining views to other tables or views, 247
    natural joins, 198
    outer joins, 183–192
        left versus right, 187
        self, 190
        three-way, 188
    self-joins, 93

specifying type of join, 84

L

last_day( ) function, 138
latin1 character set, 20
leaf nodes (B-tree indexes), 232
left outer joins, 187
left( ) function, 73
length( ) function, 119
like operator, 74
    comparing strings, 122
    regular expressions and, 76
limit clauses, 58
links, table, 51
locate( ) function, 120
locking, 218
    granularities of locks, 218
    storage engines and, 224
low-cardinality data, 233

M

matching conditions, 73
    using multiple search expressions, 75
    using regular expressions, 75
    using wildcards, 73
        example search expressions, 74
max( ) function, 145
membership conditions, 71
    generating using subqueries, 72
    using not in operator, 72
metadata, 8, 257–270
    exercises in, 270
    information included in, 257
    information_schema objects, 258–262
    publishing by database servers, 258
    using in deployment verification, 265
    using in dynamic SQL generation, 266–270
    using in schema generation scripts, 262–
        265
min( ) function, 145
mod( ) function, 127
mode, checking and changing for MySQL,
    115
modulo operator, 127
multibyte character sets, 19
multiparent hierarchy, 3
MySQL, x
    bank schema (example), 38–40

constraint generation, indexes and, 239
creating a sample database, 16
data types, 18
    character, 18
    numeric, 21
downloading and installing MySQL 6.0
        server, 15
dynamic SQL execution, 267
except operator and, 107
grouping, with cube option not supported,
    154
if( ) function, 204
indexes, 229
information_schema database, 258
intersect operator and, 106
loading time zone data, 132
locking, 218
mysql command-line tool, 10
overview of, 12
populating and modifying tables, 30–36
    deleting data, 35
    inserting data, 31–35
    updating data, 35
set operation precedence, 111
specifying join order, 90
storage engines, 223
table creation, 25–30
    building SQL schema statements, 27
time zone settings, 131
transactions
    disabling auto-commit mode, 221
    error message for deadlock, 222
    starting, 221
updatable views, 251
mysql command-line tool
    --xml option, 34
    result sets returned by, 42
    running create table statement, 28
    using, 17

N

natural joins, 198
natural key, 5
network database systems, 3
non-equi-joins, 94
noncorrelated subqueries, 158–167
    multiple-column, 165
    multiple-row, single-column, 160–165
        using all operator, 163

        using any operator, 165
nonprocedural languages, 9
normalization
    defined, 6
    table design in MySQL and, 26
not exists operator, 170, 171
not in operator, 72
    <> all versus, 163
    using in subquery, 162
not operator, using with filter conditions, 65
null values, 29
    comparisons with not in and <> all
        operators, 163
    filtering, 76
    handling using case expressions, 214
    handling when performing aggregations,
        149
    subqueries generating data for columns
        allowing null values, 180
numeric data, 126–130
    controlling number precision, 128
    converting strings to, using cast( ) function,
        141
    performing arithmetic functions, 126
    signed data, 130
numeric data types, MySQL, 21
    conditions specifying ranges of numbers,
        70
    floating-point types, 22
    integer types, 22

O

on subclause of from clause, 83
    ANSI join syntax in, 86
    joining three or more tables, 88
open source database servers, 13
operators, 63
    (see also names of individual operators)
    and, or operators in filter conditions, 63
    in filter conditions, 66
optimizers, 9
or operator
    in filter conditions, 64
    three-condition evaluation using and, or,
        64
    three-condition evaluation using and, or,
        and not, 65
    two-condition evaluation using, 64
    using in select statement where clause, 54

Oracle Database, 12
    bitmap indexes, 233
    chr( ) function, 117
    concatenation operator (||), 118, 124
    constraint generation, indexes and, 239
    decode( ) function, 204
    drop index command, 230
    dynamic SQL execution, 266
    from clauses and, 18
    generating current date or time from strings,
        137
    grouping, with cube option, 154
    inserting and updating data through views,
        254
    instr( ) function, 120
    locking, 218
    metadata, 258
    minus operator, 107
    mod( ) function, 127
    new_time( ) function, 139
    power( ) function, 127
    replace( ) function, 125
    rollups, 153
    sequences, 31
    specifying join order, 90
    starting transactions, 221
    substr( ) function, 125
    subtracting dates, 141
    text indexes and search tools, 234
    time zone settings, 132
    to_date( ) function, 136
Oracle PL/SQL language, 266
order by clauses, 34
    in select statements, 55–59
        ascending and descending sort order,
            57
        sorting via expressions, 58
        sorting via numeric placeholders, 59
    sorting compound query results, 108
outer joins, 183–192
    left versus right, 187
    self, 190
    three-way, 188
    using subqueries instead of, 206

P

page locks, 218
Perl, SQL integration toolkit, 10
PL/SQL language, 266

position( ) function, 119
PostgreSQL, 13
pow( ) or power( ) function, 127
precedence of set operations, 109
prepare statements, 267
primary key constraints, 238
    getting information about, 264
    query returning number of, 265
primary keys, 5
    defined, 7
    generating numeric values for, 31
    nonunique key values causing SQL
        statement errors, 36
procedural languages, 9
programming languages, ix
    integrating SQL with, 9
    nonprocedural, 9
Python, SQL integration toolkit, 10

Q

queries, 41
    (see also select statements)
    tuning, 236
query optimizers, 42
quote( ) function, 116

R

range conditions, 68
    correlated subqueries in, 168
    string ranges, 70
    using between operator, 69
read locks, 218
regexp operator, 76
    using in string comparisons, 122
regular expressions, 122
    (see also regexp operator)
    using to build search expressions, 75
relational databases
    mature, popular commercial products, 12
    relational model, 4
replace( ) function, 125
result sets
    defined, 7
    intermediate, 90
    returned by mysql tool (example), 42
    returned by subqueries, 157
    transformations performed with case
        expressions, 208

current_timestamp( ) function, 137
datediff( ) function, 141
datepart( ) function, 140
drop index command, 230
dynamic SQL execution, 266
generating XML from query output, 35
grouping, with cube option, 154
inserting and updating data through views, 254
len( ) function, 119
locking, 218
metadata and information_schema, 258
modulo operator (%), 127
power( ) function, 127
replace( ) and stuff( ) functions, 125
specifying join order, 90
transactions
    disabling auto-commit mode, 221
    savepoints, 225
    starting, 221
SQL92 version of ANSI SQL standard, 86
start transaction command, 221
statement classes, 7
statement scope, 157
statements
    case expressions in, 205
    clauses, 11
    data and schema, x
    dynamic SQL execution in MySQL, 267
    errors in data statements, 36
    examples of, 10
    for updatable views, 251
storage engines
    choosing, 223
    locking and, 218
STRAIGHT_JOIN keyword, 90
strcmp( ) function, 120
strict mode, 115
strings
    character data types in string data, 113
    conversion to datetimes in MySQL, 37
    converting to number using cast( ) function, 141
    generating string data, 114–118
        single quotes in strings, 116
        special characters in strings, 117
    manipulating, 119–125
        string functions returning numbers, 119–122

string functions returning strings, 123–125
partial string matches, 73
ranges of, 70
representing temporal data, 132
    date format components, 133
    functions generating dates, 135
    required date components, 134
    strig-to-date conversions, 135
SQL statements submitted to database server as, 266
temporal functions that return, 139
str_to_date( ) function, 135
subqueries, 157–182
    correlated, 167–171
        data manipulation with, 170
        using with exists operator, 169
    defined, 157
    exercises in, 181
    generating membership conditions, 72
    in statements for updatable views, 251
    noncorrelated, 159
        multicolumn, 165
        multiple-row, single-column, 160–165
    summary of types, 181
    tables generated by, 49
    types of, 158
    using as data source, 172–177
        data fabrication with subqueries, 173
        task-oriented subqueries, 175
    using as tables, 90
    using in filter conditions, 177
    using instead of outer joins, 206
    using to generate expressions, 177
    using to limit number of joins, 189
substring( ) function, 125
substrings
    extracting from strings, 125
    locating within strings, 119
sum( ) function, 146, 210
surrogate key, 5
Sybase Adaptive Server, 12
system catalog, 258

T

table aliases, 52
    queries requiring, 93
table locks, 218
tables, 4